Functional Materials Based on or Derived from Coordination Compounds for a Sustainable Society

Functional Materials Based on or Derived from Coordination Compounds for a Sustainable Society

Editors

Wu Tian
Hubei University of Education, China

Li Lu
National University of Singapore, Singapore

World Scientific

NEW JERSEY · LONDON · SINGAPORE · BEIJING · SHANGHAI · HONG KONG · TAIPEI · CHENNAI

Published by

World Scientific Publishing Co. Pte. Ltd.

5 Toh Tuck Link, Singapore 596224

USA office: 27 Warren Street, Suite 401-402, Hackensack, NJ 07601

UK office: 57 Shelton Street, Covent Garden, London WC2H 9HE

British Library Cataloguing-in-Publication Data
A catalogue record for this book is available from the British Library.

FUNCTIONAL MATERIALS BASED ON OR DERIVED FROM COORDINATION COMPOUNDS FOR A SUSTAINABLE SOCIETY

Copyright © 2025 by World Scientific Publishing Co. Pte. Ltd.

All rights reserved. This book, or parts thereof, may not be reproduced in any form or by any means, electronic or mechanical, including photocopying, recording or any information storage and retrieval system now known or to be invented, without written permission from the publisher.

For photocopying of material in this volume, please pay a copying fee through the Copyright Clearance Center, Inc., 222 Rosewood Drive, Danvers, MA 01923, USA. In this case permission to photocopy is not required from the publisher.

ISBN 978-981-98-0576-1 (hardcover)
ISBN 978-981-98-0577-8 (ebook for institutions)
ISBN 978-981-98-0578-5 (ebook for individuals)

For any available supplementary material, please visit
https://www.worldscientific.com/worldscibooks/10.1142/14114#t=suppl

Desk Editor: Rhaimie Wahap

Typeset by Stallion Press
Email: enquiries@stallionpress.com

Printed in Singapore

Contents

Preface ix

Chapter 1 Ethanol monitoring gas sensor based on Co_2P nanomaterials by a facile hydrothermal approach 1

Zhang Li, Luo Qingxia, Song Chengwen and Zhang Xiaoxing

Chapter 2 Preparation and upconversion luminescence properties of $Yb^{3+}/Tb^{3+}/Ho^{3+}$ tri-doped phosphate glasses 15

Tao Zhang, Danyi Zhang, Pei-an Wang and Caixia Cui

Chapter 3 PVA-assisted synthesis of Zn_2SiO_4: Mn^{2+} nanoparticles: Enhanced luminescence properties and applications 35

Huiya Li, Yaoyao Li, Xiaomeng Wang, Zicheng Zhang, Dejia Liu, Haiyun Ma, Hongqiang Qu, Jianzhong Xu, Yuanyuan Han and Liyong Wang

Chapter 4 Graphene dot-embedded porous WO3 photoanode for highly efficient photoelectrochemical water splitting 53

Zhenyan Xu, Xianfeng Zhao, Xiao Xu Yan, Huidan Lu and Yongping Liu

vi *Contents*

Chapter 5 High activity and sinter-resistance of Ni@silicalite-1 catalyst for dry reforming of methane 75

Chunlan Han, Xiaoxiao Zhu and Xinping Wang

Chapter 6 On the successful growth of bulk gallium oxide crystals by the EFG (Stepanov) method 87

Dmitrii Andreevich Bauman, Dmitrii Iurevich Panov, Vladislav Alekseevich Spiridonov, Arina Valerievna Kremleva and Alexey Evgenievich Romanov

Chapter 7 Resistive switching properties of CdTe/CdSe core–shell quantum dots incorporated organic cow milk for memory application 97

Zolile Wiseman Dlamini and Wendy Setlalentoa, Sreedevi Vallabhapurapu, Tebogo Sfiso Mahule and Vijaya Srinivasu Vallabhapurapu, Olamide Abiodun Daramola and Potlaki Foster Tseki, Xavier Siwe-Noundou and Rui Werner Macedo Krause

Chapter 8 Molecular simulation of (Al–Ga) surface garnished with chromium metal for organic material detecting: A DFT study 123

Fatemeh Mollaamin and Majid Monajjemi

Chapter 9 Photothermal insulation mechanism of submicron ITO hollow particles in PVDF film 149

Hedong Li, Peihu Shen, Zizheng He, Yang Xu and Minjia Wang

Chapter 10 Effect of silica morphology on rheological properties and stability of magnetorheological fluid 163

Bingsan Chen, Minghan Yang, Chenglong Fan, Xiaoyu Yan, Yongchao Xu and Chunyu Li

Contents vii

Chapter 11 Porous pyroelectric material for waste
heat harvesting   185

*Qingping Wang, Haifeng Luo, Zhanxiang Xu
and Tian Wu*

Chapter 12 Preparation of Mo^{6+}, Gd^{3+}-doped TiO_2
nanotube arrays and study of their organization
and photocatalytic properties   199

*Chaoqian Qin, Jinghong Du, Jiaxing Chen,
Jiarui Yang and Shengyang He*

Chapter 13 Preparation and photocatalytic degradation
performance of AgI/4A molecular sieves   221

Yaqian Duan and Jianping Gao

Preface

Climate changes particularly in recent decades which has largely impacted our environments has internationally been considered and recognized a major global issue. Polar ice shields melting causes sea level rise, extreme weathers such as rainfall and heat waves often result in floodings and wildfires, and air and water pollutions impact habitats' life. Therefore, it is extremely important and urgent to develop necessary policies and utilize different technologies to prevent our earth from further deterioration.

Functional materials are a current topic of interest for environmental management in the context of clean water, pollution risk assessment, CO_2 reduction, cleaner energy generation, and green fuel production, among others. Advanced functional materials encompass a vast range of hybrid and nanomaterials, including metal oxides, phosphides, graphene, carbon nitride, semiconductors, polymers, ion exchange resins, quantum dots, bi- and trimetallic nanoparticles, and ceramics. These multifunctional materials can act as sensors for heavy metals or organic pollutants and thus assist in pollution risk assessment and, at the same time, they can be explored on the basis of their adsorption and photocatalytic nature for the remediation of environment contaminants. The combination of the above materials has led to designing a new class of materials known as composites, where such materials possess multiple applications with superior properties and improved stability. The fabrication of composite materials with desired properties and applications is challenging task for a material researcher.

x *Preface*

With this aim in mind, this publication has 13 chapters dedicated to preserving a green environment using novel and advanced functional materials in the areas of detection of gases,[1,2] green energy,[3] cost-effective and environmental friendly synthesis technologies,[4-6] photocatalytic-assisted degradation of pollutants,[7,8] waste harvesting,[9] new materials for luminescence applications,[10,11] and photothermal insulation and rheological properties of functional materials.[12,13]

This publication will make a substantial contribution to the development of functional materials for green environment. I would like to take this opportunity to acknowledge contributions of all authors who have developed various technologies to protect our green environment.

References

1. Ethanol Monitoring Gas Sensor Based on Co_2P Nanomaterials by a Facile Hydrothermal Approach, https://doi.org/10.1142/S1793604723400192.
2. Molecular Simulation of (Al–Ga) Surface Garnished with Chromium Metal for Organic Material Detecting: A DFT Study, https://doi.org/10.1142/S1793604723400283.
3. Graphene Dots Embedded Porous WO_3 Photoanode for High Efficient Photoelectrochemical Water Splitting, https://doi.org/10.1142/S1793604723400246.
4. High Activity and Sinter-resistance of Ni@silicalite-1 catalyst for Dry Reforming of Methane, https://doi.org/10.1142/S1793604723400258.
5. Resistive Switching Properties of CdTe/CdSe Core-Shell Quantum Dots incorporated Organic Cow Milk for Memory Application, https://doi.org/10.1142/S1793604723400271.
6. On the Successful Growth of Bulk Gallium Oxide Crystals by the EFG (Stepanov) Method, https://doi.org/10.1142/S179360472340026X.
7. Preparation and Photocatalytic Degradation Performance of AgI/4A Molecular Sieves, https://doi.org/10.1142/S1793604723400398.
8. Preparation of Mo^{6+}, Gd^{3+} doped TiO_2 Nanotube Arrays and Study of Their Organization and Photocatalytic Properties, https://doi.org/10.1142/S1793604723400374.

9. Porous Pyroelectric Material for Waste Heat Harvesting, https://doi.org/10.1142/S1793604723400362.
10. PVA-assisted Synthesis of Zn_2SiO_4: Mn^{2+} Nanoparticles: Enhanced Luminescence Properties and Applications, https://doi.org/10.1142/S1793604723400222.
11. Preparation and Upconversion Luminescence Properties of $Yb^{3+}/Tb^{3+}/Ho^{3+}$ Tri-doped Phosphate Glasses, https://doi.org/10.1142/S1793604723400209.
12. Effect of Silica Morphology on Rheological Properties and Stability of Magnetorheological Fluid, https://doi.org/10.1142/S1793604723400301.
13. Photothermal Insulation Mechanism of Submicron ITO Hollow Particles in PVDF Film, https://doi.org/10.1142/S1793604723400295.

Tian Wu
Hubei University of Education, China

Lu Li
National University of Singapore

Chapter 1

Ethanol monitoring gas sensor based on Co_2P nanomaterials by a facile hydrothermal approach

Zhang Li[*,‡], Luo Qingxia[†], Song Chengwen[†] and Zhang Xiaoxing[†]

[*]*College of Environmental Science and Engineering Dalian Maritime University, Dalian, 1 Linghai Road/116026, P. R. China*
[†]*Dalian, 1 Linghai Road/116026, P. R. China*
[‡]*zl780725@163.com*

Co_2P nanocomposites were successfully synthesized via a facile hydrothermal method. The microstructure, morphology and elemental composition were examined by XRD, SEM, TEM and XPS. The effects of synthesis temperature and reaction time on the sensing properties of Co_2P nanocomposites were analyzed. The Co_2P sensor at 200°C and 3 h reaction condition exhibited optimum sensitivity toward 100 ppm ethanol. In comparison with other gases, ethanol possesses good selective characteristics at optimal operating temperature of 160°C, which greatly reduce energy consumption. The above results showed that Co_2P nanocomposites have the potential application as an effective sensor for ethanol detection.

Keywords: Synthesis, Co_2P, ethanol, gas-sensor, detection.

[‡]Corresponding author.

To cite this article, please refer to its earlier version published in the Functional Materials Letters, Volume 16(7), 2340019 (2023), DOI: 10.1142/S1793604723400192.

1. Introduction

In the modern society, the rapid development of oil industry and chemical industry inevitably produces or exhausts various toxic, flammable, explosive and harmful volatile organic compounds including benzene, toluene, acetone, formaldehyde, phenol, ethanol and so forth, which usually cause a great threat to the environmental and human health problems.[1,2] Among them, ethanol has received widespread attention as ethanol is the key industrial raw material used in various fields such as alcoholic drinks, cosmetics, lacquers, and PPCPs (Pharmaceuticals and Personal Care Products).[3-6] In order to detect ethanol leakage or drunk driving, it is an significant need to develop a fast, selective and sensitive method for ethanol detection. The accurate gas detection via analytical apparatuses, including chromatography and mass spectrometry, is inconvenient and expensive for effective measurements.[7] Therefore, the low-cost, convenient, selective and sensitive sensors for detection of ethanol are urgent and important. Up to now, a variety of ethanol detection gas sensors have been reported, each with its own characteristics, such as fast-response, simplicity, cost efficiency, outstanding selectivity, high sensitivity and excellent reproducibility.[8]

More and more available gas sensors widely reported by many research groups, most semiconductor gas sensors using metal oxides (such as ZnO,[5] SnO_2,[3] Co_3O_4,[9] Fe_3O_4,[10] CuO,[11] TiO_2,[12] WO_3[13]), metal sulfifides (such as ZnS,[14] CuS,[15] PbS,[16] MoS_2,[17] CdS^1), and mental selenides (such as WSe_2,[2] $MoSe_2$,[18] $ZnSe$,[19] $SnSe_2$[20]) as sensing materials have been systematically investigated because of their lower power consumption, excellent linear response, remarkable sensing performance at room temperature.[8] In addition, the development of new gas sensing materials has attracted the attention of scholars.

Transition metal phosphides (TMPs) are a sort of important functional materials obtained by the reaction between transition metal and phosphorus with instinct and superior electrical conductivity, thermal conductivity, durability and high catalytic activity, which have attracted intense attention and been used in catalysis,[21,22] energy storage,[23] electrocatalysts for water splitting,[24] sensing[25] and other real-life fields.[26] In terms of electrochemical sensor design, Co_2P materials are used in the detection of glucose,[27,28] especially Co_2P encapsulated

N, P dual doped carbon nanotubes (Co_2P/NPCNTs) were synthesized by Das *et al.* and for the first time demonstrated to be an electrochemical sensor for the enzymeless detection of glucose.[27] Co_2P can also detect dopamine[29] and some small molecules.[30 32] For example, Wang *et al.* found that synthesized C/Co_2P/Pd sensor demonstrates better sensing response than that of the C/Co_2P and C/Pd sensor for HCHO detection.[31] Cobalt phosphide nanoparticles were prepared by Yin *et al.*, which exhibited their potential application as a sensitive platform for H_2O_2 detection.[32] A CoxP (mixture of CoP and Co_2P) embedded within nitrogen-doped porous carbon microspheres was easily made to determine 4-nitrophenol.[30] Accordingly, Co_2P nanocomposite as a promising sensitive sensor is studied to detect some VOCs.

Herein, in our work, we report the synthesis of Co_2P gas sensors via a simple and mild one-step hydrothermal method utilizing cobalt chloride and sodium hypophosphite as phosphorus source. The structure and morphology of prepared Co_2P were examined by XRD, SEM, TEM, XPS and the gas sensing toward ethanol of Co_2P nanocomposites was investigated systematically by optimizing reaction temperature and reaction time. We found that the synthesized Co_2P in our study showed good sensitivity and selectivity to ethanol compared to some other gases, which has the potential to be applied as an effective sensor for ethanol detection. To our knowledge, Co_2P as gas sensors has been rarely used to research ethanol detection. Besides that, the corresponding ethanol sensing mechanism of Co_2P is systematically discussed.

2. Experimental

2.1. *Fabrication of Co_2P nanomaterials*

About 1 mmol of $CoCl_2 \cdot 6H_2O$, 15 mmol $NaH_2PO_2 \cdot H_2O$ and 0.1 g PVP were respectively dissolved into 40 mL deionized water under stirring with 20 r/s speed. At room temperature, the stirring was carried out for at least 30 min until the synthesized solution became clear and pink. Then the precursor is moved into a Teflon-lined stainless steel autoclave of 100 mL and heated to 180°C, 190°C, 200°C for 3 h, 5 h, 7 h, respectively. The prepared solution was cooled naturally to room temperature and filtered. In order to remove any presented impurities, the obtained products were centrifuged and washed respectively by deionized water and ethanol for three times, and then dried at 60°C for 3 h.

2.2. Nanomaterials characterizations

The crystallographic structure of Co_2P samples with different reaction temperatures was identified by X-ray diffractometer (XRD, Cu Kα radiation, D/Max-2400). The morphologies and internal structure of Co_2P nanomaterials were characterized by a scanning electron microscope (SEM) (Philips XL 30 FEG) and transmission electron microscopy (TEM, FEI/Philips Techal 12 BioTWIN). The chemical element composition of the Co_2P samples was identified by XPS on a thermo SCIENTIFIC ESCALAB 250 spectrometer.

3. Results and Discussion

3.1. Morphology and structure of Co_2P nanomaterials at different reaction temperatures

As presented in Fig. 1, the phase structures and purity of the Co_2P nanomaterials synthesized at 180, 190 and 200°C with reaction

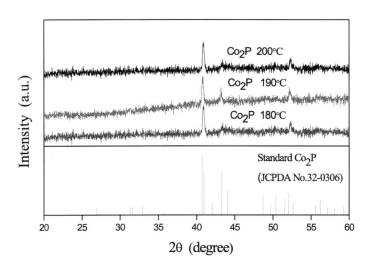

Fig. 1. The XRD patterns of Co_2P samples.

time 3 h were analyzed by X-ray diffraction (XRD) patterns. As observed, the characteristic diffraction peaks at 2θ values are $40.7°$, $40.9°$, $43.3°$, $44.1°$, $52.1°$, $56.2°$ which could be indexed to (121), (201), (211), (130), (002), (320) crystalline planes of Co_2P, which match well with standard orthorhombic structure of Co_2P (JCPDS Card. No. 32-0306).[31,33 35] In addition, no other diffraction peaks are found, which is suggested the high crystallinity of pure Co_2P nonomaterials were prepared.[36]

X-ray photoelectron spectroscopy (XPS) was used to analyze the composition and elemental binding state of the sample, which is presented in Fig. 2. The used Co_2P sample underwent the same synthetization and thermal treatment as the sensors. Figure 2(a) shows the survey spectrum, which indicates that the prepared sample is composed of Co and P elements. In the high-resolution spectrum of Co 2p (Fig. 2(b)), two peaks at the binding energy of 777.8 eV, 793.1 eV could be attributed to Co $2p_{3/2}$ and Co $2p_{1/2}$. Figure 2(c) presents the P 2p spectrum, the peaks appearing at 128.9 eV and 130.0 eV could be assigned to the P $2p_{3/2}$ and P $2p_{1/2}$, respectively.[37]

Figure 3 depicts the morphologies of the resultant Co_2P samples at three synthesis temperatures of 180°C, 190°C and 200°C. Figures 3(a) and 3(b) show the flower-like morphologies of Co_2P under 1 μm and 5 μm, when the reaction temperature was fixed at 180°C. The flower-like morphologies had no smooth surface and looked like they were composed of many small particles. When the reaction temperature reached 190°C, the particles of the flower-like nanostructures split (Figs. 3(c) and 3(d)). By continuing to increase the temperature to 200°C, the smaller diameter and more broken, sharpen fragments of nanostructures are shown in Figs. 3(e) and 3(f).[15] The increasing surface area of Co_2P is helpful to improve the adsorption capacity and increase the performance of Co_2P sensing. Therefore, the Co_2P materials synthesized at 200°C may have better gas sensing properties, which is consistent with the gas sensitivity.

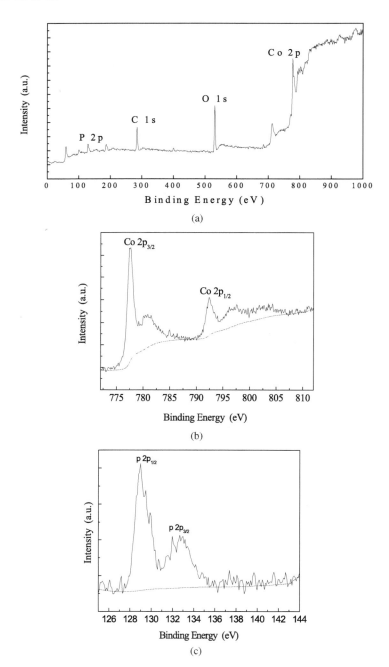

Fig. 2. The XPS spectra of the Co$_2$P sample synthesized at 200°C, (a) wide-scan spectrum, (b) Co 2p spectrum and (c) p 2p spectrum.

Ethanol monitoring gas sensor 7

Fig. 3. The SEM images of Co₂P synthesized at (a, b) 180°C, (c, d) 190°C, (e, f) 200°C under 1 μm and 5 μm, respectively.

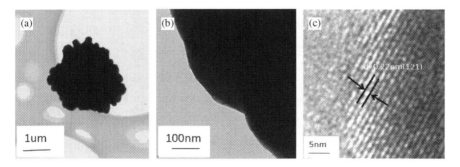

Fig. 4. The TEM image of Co₂P nanocomposites.

To gain more morphological information and insight into the microstructure of the Co₂P, TEM images are shown in Fig. 4. It indicates that the morphology of Co₂P is spherical and its diameter is about 2 μm, which is consistent with SEM. The HRTEM images of Co₂P show that the prepared Co₂P material has the clear crystal lattice (lattice distance of 0.22 nm matches the (121) facets of

3.2. Gas sensing of Co_2P

It is well known that the operating temperature determined largely the gas sensing performance of gas sensor. Figure 5 illustrates the sensing performance of synthesized Co_2P at different reaction temperatures (180°C, 190°C, 200°C), with Co_2P exposed to working temperature ranging from 80°C, 100°C, 160°C, 200°C to 240°C and toward 100 ppm ethanol. Figure 6 presents the sensing performance of Co_2P at different reaction times (3 h, 5 h, 7 h), which were exposed to the same working temperature and sensing gas as in Fig. 5. We can reveal from Figs. 5 and 6 that the optimum reaction conditions are 200°C synthesis temperature and 3 h reaction time, which demonstrated the highest gas response. It is noteworthy that the sensing

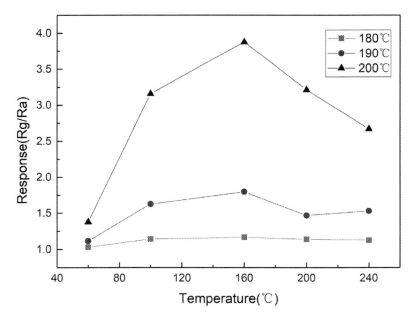

Fig. 5. Response of the Co_2P sensor at different synthesis temperatures with 3 h reaction time.

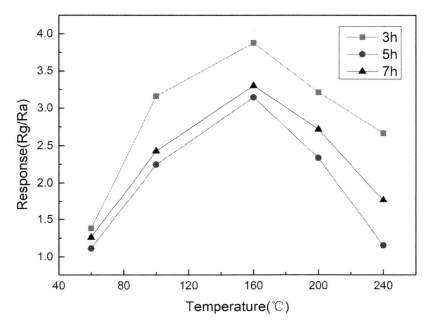

Fig. 6. Response of the Co$_2$P sensor at different synthesis times with 200°C reaction temperature.

performance of synthesized Co$_2$P increased and then reached the maximum sensitivity when the working temperature was 160°C. At the best working temperature, the highest gas response may be attributed to enough energy for ethanol molecules to react with oxygen species O_2^-, O^-, O^{2-}. As far as we know, energy source is the essential and prominent resource for human survival and development in the current world. In our study, working temperature of Co$_2$P is 160°C, which is a relative low working temperature for gas sensors, and significantly reduces the energy consumption. To the best of our knowledge, synthesized Co$_2$P in our study is advantageous for gas-sensitive performance.

Figure 7 shows the sensing measurements are further inspected for various concentrations of alcohol at its maximum working temperature (160°C). When the Co$_2$P sensor was contacted ethanol with a relatively low concentration (less than 100 ppm), a rapidly

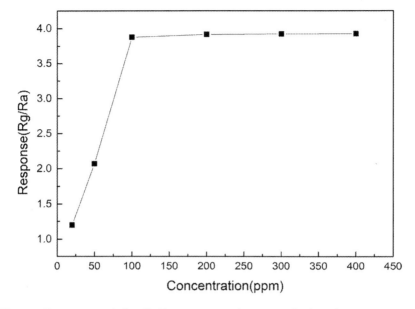

Fig. 7. Responses of the Co$_2$P sensor as a function of ethanol concentration.

gas-sensitive response with a linear trend was presented, which is shown that Co$_2$P sensor is suitable for detecting low concentration ethanol leakage. With the improvement of the gas concentration, the sensor response increased slowly and displayed a tendency to attain its own saturation lever, which was attributed to the limited amount of adsorption of ethanol molecules and hardly transferred electrons.

Figure 8 presents the response and recovery characters of Co$_2$P sensor towards different ethanol concentrations (20, 50, 100, 200, 300 ppm). When the ethanol gas was injected into the gas sensing measurement system, the response of Co$_2$P sensor improved rapidly, with about 1 s response time and less than 20 s recovery time. It is well known that the response and recovery time of gas sensors are kept within 30 s, indicating the sensor has high practical value and good application prospect. So, Co$_2$P can be used to detect ethanol as a good gas sensor.

As shown in Fig. 9, the sensitivity of Co$_2$P sensor to methanol, benzene, formaldehyde, dichloromethane, ammonia, acetone, *n*-hexane

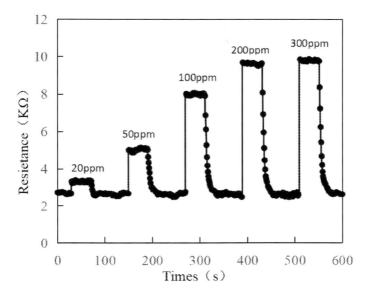

Fig. 8. The response and recovery curve of the Co$_2$P sensor towards ethanol.

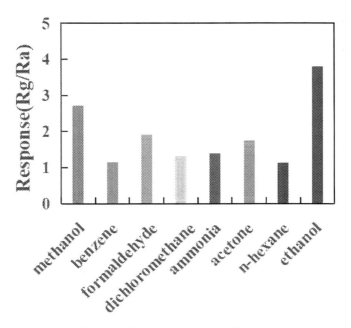

Fig. 9. Selectivity of the Co$_2$P sensor to different gases at 100 ppm concentration.

and ethanol gases was measured at a concentration of 100 ppm. Obviously, the significant response to the alcohol showed slightly less sensitivity to methanol, and weak sensitivity to other gases, especially benzene and *n*-hexane. Therefore, it could be concluded that the optimum and excellent selectivity of ethanol gas suggesting Co_2P sensor showed a great potential in its utility for ethanol detection.

The physical and chemical adsorptions on the Co_2P material surface occur when it is contacted with the reducing gases of ethanol. The gas-sensitive Co_2P material adsorbs the oxygen, then captures free electrons and forms oxygen species (O_2^-, O^-, O^{2-}), which is described in the following equations:

$$O_2(gas) \rightarrow O_2(ads) \tag{1}$$
$$O_2(ads) + e^- \rightarrow O_2^- (ads) \tag{2}$$
$$O_2^- (ads) + e^- \rightarrow 2O^- (ads) \tag{3}$$
$$O^- (ads) + e^- \rightarrow O^{2-} (ads) \tag{4}$$

At the same time, the ionization of the ethanol is accompanied by the reaction with oxygen species (O^-). The related reaction process is presented by the following equations:

$$C_2H_5OH(gas) \rightarrow C_2H_5OH(ads) \tag{5}$$
$$C_2H_5OH(ads) + 6O^- (ads) \rightarrow 2CO_2(gas) + 3H_2O(gas) + 6e^- \tag{6}$$
$$C_2H_5OH(ads) + 6O^{2-} (ads) \rightarrow 2CO_2(gas) + 3H_2O(gas) + 12e^- \tag{7}$$

4. Conclusions

In summary, the novel Co_2P gas-sensing materials were synthesized by a one-step hydrothermal method, which possessed ball-like nanostructures. The sensor of Co_2P synthesized at 200°C and 3 h reaction time showed excellent sensitivity to alcohol gas against many other reducing gases including methanol, benzene, ammonia, dichloromethane, formaldehyde, formaldehyde, acetone, and *n*-hexane. It is a remarkable fact that Co_2P has lower operating temperature (160°C), excellent selectivity, sensitivity and response time, so Co_2P can be used as sensing material for alcohol detection.

References

1. X. Q. Yan *et al.*, *ACS Omega* **7**, 1468 (2022).
2. A. Hussain *et al.*, *Physica E* **147**, 1 (2023).
3. C. S. Reddy *et al.*, *J. Alloys Compd.* **813**, 1 (2020).
4. X. Yang *et al.*, *Actuat. B-Chem.* **282**, 339 (2019).
5. H. R. Madvar *et al.*, *Sensors* **23**, 365 (2023).
6. Y. Tan and J. Zhang, *Physica E* **147**, 1 (2023).
7. M. Li *et al.*, *Sci. Rep.* **3**, 1 (2013).
8. H. Y. Li *et al.*, *Chem. Soc. Rev.* **17**, 6364 (2020).
9. R. Zhang *et al.*, *Appl. Surf. Sci.* **503**, 1 (2020).
10. B. Sharma *et al.*, *Appl. Surf. Sci.* **530**, 1 (2020).
11. U. Nakate *et al.*, *Microelectron. Eng.* **251**, 1 (2022).
12. C. Duan *et al.*, *Micromachines* **14**, 156 (2023).
13. J. K. Xiao *et al.*, *Mater. Res.* **24**, 3 (2021).
14. H. Parangusan *et al.*, *Talanta* **219**, 1 (2020).
15. D. P. Wang *et al.*, *J. Mater. Eng. Perform.* **28**, 6649 (2019).
16. A. K. Mishra and S. Saha, *Int. J. Electrochem. Sci.* **15**, 11594 (2020).
17. M. Gajanan *et al.*, *J. Alloy. Comp.* **941**, 1 (2023).
18. Y. W. Luo, D. Z. Zhang and X. Fan, *IEEE SENS. J.* **15**, 13262 (2020).
19. U. Choudhari and S. Jagtap, *J. Electron. Mater.* **49**, 5903 (2020).
20. Q. N. Pan, T. T. Li and D. Z. Zhang, *Sensor Actuat. B-Chem.* **332**, 1 (2021).
21. S. Gutierrez-Rubio *et al.*, *Catal. Today* **390**, 135 (2022).
22. D. N. Liu *et al.*, *Nanotechnology* **27**, 1 (2016).
23. S. Li *et al.*, *J. Mater. Chem. A* **7**, 17386 (2019).
24. Y. Wang *et al.*, *Nano. Today* **15**, 26 (2017).
25. T. Chen *et al.*, *Anal. Chem* **88**, 7885 (2016).
26. L. Zhang *et al.*, *Microchim. Acta* **186**, 309 (2019).
27. D. Das *et al.*, *Green Chem.* **19**, 1327 (2017).
28. X. Y. Lu *et al.*, *Chinese J. Inorg. Chem.* **36**, 1675 (2020).
29. M. We *et al.*, *J. Mater. Sci* **56**, 6401 (2021).
30. L. L. Xiao, R. Y. Xu and F. Wang., *Talanta* **179**, 448 (2018).
31. H. Wang *et al.*, *J. Chem.* **2017**, 1 (2017).
32. D. H. Yin *et al.*, *Nanoscale Res. Lett.* **16**, 1 (2021).
33. C. Zhang *et al.*, *Appl. Surf. Sci.* **459**, 665 (2018).
34. Y. He *et al.*, *J. Power Sources* **402**, 345 (2018).
35. J. Li *et al.*, *Int. J. Hydrogen Energ.* **43**, 1365 (2018).

36. D. Das and K. K. Nanda, *Nano Energy* **30**, 303 (2016).
37. M. Cheng *et al.*, *Nanoscale* **37**, 1 (2017).
38. Z. P. Huang *et al.*, *Nano Energy* **9**, 373 (2014).
39. D. L. Jiang *et al.*, *J. Colloid Interf. Sci.* **550**, 10 (2019).
40. M. H. Chen *et al.*, *J. Power Sources*, **342**, 964 (2017).

Chapter 2

Preparation and upconversion luminescence properties of $Yb^{3+}/Tb^{3+}/Ho^{3+}$ tri-doped phosphate glasses

Tao Zhang[*,¶], Danyi Zhang[†], Pei-an Wang[‡] and Caixia Cui[§]

[*]*School of Information and Communication, Guangxi Key Laboratory of Wireless Broadband Communication and Signal Processing Guilin University of Electronic Technology, Guilin 541004, Guangxi, P. R. China*
[†]*School of Ocean Engineering, Guilin University of Electronic Technology, Beihai 536000, Guangxi, P. R. China*
[‡]*School of Information and Communication, Guilin University of Electronic Technology, Guilin 541004, Guangxi, P. R. China*
[§]*School of Economics and Management, Guilin University of Electronic Technology, Beihai 536000, Guangxi, P. R. China*
[¶]*63407583@qq.com*

In this study, Yb^{3+}/Ho^{3+}, Yb^{3+}/Tb^{3+} co-doped, and $Yb^{3+}/Ho^{3+}/Tb^{3+}$ tri-doped phosphate glasses have been prepared by the high-temperature melting method using P_2O_5 as the base material, and these phosphate glasses are characterized as non-crystalline structures by X-ray diffraction. Under 980 nm excitation, highly efficient blue (489 nm), green (545 nm), and red (661 nm) upconversion luminescences can be observed in this glass system, which are attributed to Tb^{3+}: $^5D_4 \rightarrow {}^7F_6$, Tb^{3+}: $^5D_4 \rightarrow {}^7F_5$, and

---

[¶]Corresponding author.

To cite this article, please refer to its earlier version published in the Functional Materials Letters, Volume 16(7), 2340020 (2023), DOI: 10.1142/S1793604723400209.

Ho^{3+}: $^5F_5 \rightarrow {}^5I_8$ radiative transitions, respectively. The upconversion emission intensity and excitation power dependence analysis reveal that the blue and green light emissions are three-photon processes, while red light emission is a two-photon process. White light emission can be achieved by adjusting the doping concentration of rare-earth ions in the Yb^{3+}/Ho^{3+}/Tb^{3+} tri-doped phosphate glasses. The CIE chromaticity coordinates of 2Yb^{3+}/0.1Ho^{3+}/0.4Tb^{3+} tri-doped phosphate glass (0.39, 0.39) under 980 nm excitation are relatively close to those of the standard light source (0.33, 0.33), indicating the potential application of this material in the field of illumination, such as emissive displays, fluorescent lamps and fiber lasers.

Keywords: Upconversion; Yb^{3+}/Ho^{3+}/Tb^{3+}; phosphate glass; energy transfer.

1. Introduction

Over the recent years, white light sources have attracted considerable research attention in a variety of applications, such as semiconductor lighting, solid-state 3D displays and bio-imaging.[1–3] Notably, upconversion luminescence technology is one of the effective ways to achieve white light emission. The basic essence of this technique is that the luminescent material absorbs long wavelength light and emits short wavelength light, i.e., the conversion of near-infrared (NIR) light into visible light of multiple wavelengths, which is a kind of anti-Stokes photoluminescence phenomenon.[4–7] Among all the rare-earth ions, Ho^{3+} has abundant energy level structures and is commonly used as the activating ion for the upconversion luminescence of red and green light.[8,9] On the other hand, Tb^{3+} ion has seven 4f layer electrons, and its luminescence is determined by the energy level transitions between the 4f ground state and the 4f excited state. Further, its emission spectrum shows a series of peaks in the blue, green, yellow, and red bands.[10–13] However, these two ions have a small absorption cross section for infrared light at 980 nm, resulting in a low fluorescence intensity and pumping efficiency of the fluorescent materials doped with these two activating ions, so sensitizers are needed to improve the upconversion luminescence efficiency. Since Yb^{3+} has a simple energy level structure, a long excited state lifetime, and a large absorption cross section for infrared light at 980 nm, it is a suitable candidate for sensitizing ion.

Therefore, Yb^{3+}, Ho^{3+}, and Tb^{3+} are usually simultaneously doped in the matrix material to achieve white light emission.

In lighting applications, commercial white light-emitting diodes (W-LEDs) are typically used to emit white light by utilizing GaN LED chips (blue) to excite a Ce^{3+}:YAG phosphor (yellow). Although such LEDs have the advantages of high efficiency and environmental protection,[14,15] they suffer from poor thermal stability and low emitted light uniformity, which affects their service life.[16,17] Several previous reports have suggested that the doping of rare-earth ions in glass matrix can substantially improve the luminescence efficiency and thermal stability. For example, Dy^{3+}/Tm^{3+} co-doped oxyfluoride germanate glasses,[18] $Tm^{3+}/Ce^{3+}/Tb^{3+}$ tri-doped silicate glasses,[19] Dy^{3+}/Eu^{3+} co-doped zinc–aluminium–sodium–phosphate (ZANP) glasses,[20] and $Ce^{3+}/Tb^{3+}/Eu^{3+}$ tri-doped calcium borosilicate glass have been prepared and their potential application in white light emission has been discussed.[21] However, the higher phonon energy of these glass systems results in a lower luminescence efficiency of rare-earth ions. By contrast, the phosphate glass system has the advantages of high thermal expansion coefficient, strong thermal stability, and excellent mechanical properties, and its lower phonon energy allows it to dissolve more rare-earth ions without causing serious concentration quenching.

Based on the above understandings, in this study, a new upconversion luminescent material of phosphate glass doped with rare-earth ions is prepared, and the intensity ratio of blue, green, and red light is adjusted by varying the doping concentrations of $Yb^{3+}/Ho^{3+}/Tb^{3+}$ under 980 nm excitation to realize white light emission with excellent performance.

2. Experimental Method

The composition of glasses is shown in Table 1. La_2O_3-P_2O_5 was used as the matrix material of fluorescent glass, which was doped with rare-earth oxides Yb_2O_3, Ho_2O_3, and Tb_4O_7, all with raw material purity of 4 N, and a series of phosphate glasses with various concentrations of rare-earth ions were fabricated by high-temperature melting method. The preparation process is described as follows:

18 T. Zhang et al.

Table 1. Raw material composition (mol%) of the samples.

Samples	La_2O_3	P_2O_5	Yb^{3+}	Tb^{3+}	Ho^{3+}
1	15	85	2	0.4	0.0
2	15	85	2	0.4	0.1
3	15	85	2	0.4	0.2
4	15	85	2	0.0	0.05
5	15	85	2	0.4	0.05
6	15	85	2	0.6	0.05
7	15	85	2	0.8	0.05
8	15	85	2	1.0	0.05

Firstly, each chemical reagent was weighed according to the molar percentage in a total mass of 30 g, which was then placed into an agate mortar and thoroughly ground for 10 min to obtain a uniformly mixed powder. Subsequently, the powder was poured into an alumina crucible and preheated in a heating furnace at 500°C for 1 h to ensure that the matrix material fully reacted with the rare-earth oxides. Secondly, the alumina crucible was transferred to a high-temperature furnace at 1000°C and was continuously heated for 30 min, where the raw material transformed from solid to liquid state to form a melt. It was then transferred to a high-temperature furnace at 1620°C and held for 30 min, where the melt released small bubbles to generate a clear transparent glass solution. Finally, the glass solution was poured into a graphite mold and pressed into a thin sheet of uniform thickness with a copper plate at room temperature. It was necessary to anneal the glass sample in a muffle furnace at 500°C for 3 h because the stress inside the glass could easily cause spontaneous breakage. All the glass samples were cut and polished into small pieces with dimensions of 15 mm × 15 mm × 2 mm to achieve excellent transparency for subsequent optical measurement. The structures of the rare-earth doped phosphate glass samples were characterized by X-ray diffraction (XRD; Ultima4 multipurpose diffractometer) with copper Ka radiation source to determine whether crystals were precipitated from the samples. The XRD pattern was scanned over 2θ range of 10° to 90° at a scanning rate of 5°/min. To analyze the absorption properties of fluorescent glasses at different wavelengths, an Agilent 3600 plus ultraviolet (UV) spectrophotometer was utilized to obtain the absorption spectrum of the samples in the wavelength range of 200–1200 nm. An F-4000 fluorescence spectrometer was used to obtain the upconversion emission spectrum and fluorescence lifetime of the glasses, where the excitation source was a 980 nm semiconductor laser. The CIE chromaticity coordinates of $Yb^{3+}/Tb^{3+}/Ho^{3+}$ tri-doped phosphate glasses were calculated according to the 1931 CIE standard. All these measurements were performed at room temperature.

3. Results and Discussion

The XRD patterns of the glass samples are shown in Fig. 1. The three phosphate glass samples doped with rare-earth ions show

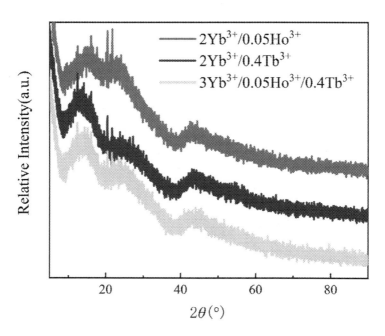

Fig. 1. The XRD patterns of phosphate glass samples #1, #4, and #5.

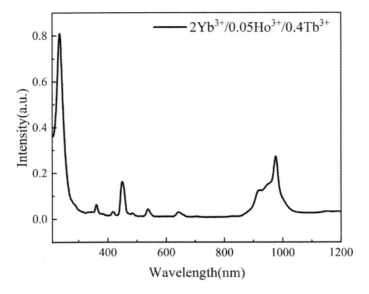

Fig. 2. Absorption spectrum of phosphate glass sample #5.

diffused humps, and no sharp diffraction peaks are observed, indicating that all the samples have short-range ordered and long-range disordered amorphous structures, confirming that the prepared materials are typical glass materials.

Figure 2 shows the absorption spectrum of $Yb^{3+}/Ho^{3+}/Tb^{3+}$ tri-doped phosphate glass, and eight strong absorption peaks are observed at 361, 379, 417, 450, 485, 538, 643, and 977 nm. The peaks at 361, 417, 450, 538, and 643 nm are attributed to the absorption of Ho^{3+} ions, while the peak at 379 nm corresponds to the absorption of Tb^{3+} ions. Further, the broad absorption band at 977 nm mainly corresponds to the multiplet state absorption of Yb^{3+} ions, and the maximum absorption cross section of Yb^{3+} ions is observed near 980 nm. Furthermore, the absorption peak at 485 nm is ascribed to the combined absorption effect of Tb^{3+} and Ho^{3+}.

Figure 3 depicts the emission spectrum of $2Yb^{3+}/0.05Ho^{3+}$ co-doped phosphate glass excited by a 980 nm laser, and the measured wavelength range is 450–700 nm. The green emission peak is roughly located at 545 nm, which is caused by the $^5S_2/^5F_4 \rightarrow {}^5I_8$

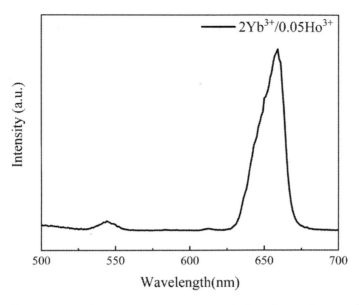

Fig. 3. Upconversion luminescence spectrum of phosphate glass sample #5 excited at 980 nm.

energy level transition of Ho^{3+}, while the red light at 661 nm originates from the $^5F_5 \rightarrow {}^5I_8$ transition of Ho^{3+}.

Figure 4 shows the upconversion emission spectrum of phosphate glass co-doped with 2Yb^{3+}/0.4Tb^{3+} under 980 nm excitation, and the measured wavelength range is 350–700 nm. It can be seen that there are seven obvious emission peaks, among which the peaks at 381, 415, and 437 nm correspond to the $^5D_3 \rightarrow {}^7F_J$ energy level transition (J = 6, 5, 4), while the peaks at 489, 545, 588, and 623 nm correspond to the $^5D_4 \rightarrow {}^7F_J$ transition (J = 6, 5, 4, 3), respectively.

Figure 5 shows the upconversion fluorescence spectrum of 2Yb^{3+}/0.4Tb^{3+}/xHo^{3+} (x = 0.05, 0.1, 0.2) tri-doped phosphate glasses under 980 nm excitation. It is clear that the blue light emission is exclusively related to Tb^{3+}, the green light originates from the combined emission of Ho^{3+} and Tb^{3+}, and the red light emission is mainly from Ho^{3+}.

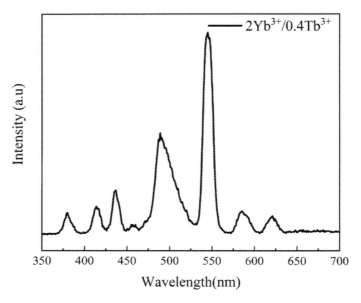

Fig. 4. Upconversion luminescence spectrum of phosphate glass sample #1 excited by 980 nm laser.

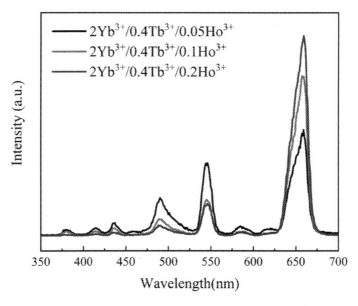

Fig. 5. Upconversion luminescence spectra of phosphate glass samples #2, #3, and #5 under 980 nm laser excitation.

Notably, although the green light is emitted by both Ho^{3+} and Tb^{3+}, it mainly originates from the $^5D_4 \rightarrow {}^7F_5$ energy level transition of Tb^{3+} ions, so the intensity of green light decreases with the increase in the Ho^{3+} concentration. On the other hand, the blue light emission is only related to Tb^{3+}, so the increase in the Ho^{3+} concentration reduces the efficiency of energy transfer from Yb^{3+} to Tb^{3+}, causing a continuous decrease in the blue light intensity.

By investigating the effect of Tb^{3+} ion concentration on the upconversion luminescence properties of Yb^{3+}/Ho^{3+}/Tb^{3+} tri-doped phosphate glasses, the emission spectra of phosphate glasses doped with different concentrations of Tb^{3+} ions were tested under 980 nm laser excitation, and the results are shown in Fig. 6. It is evident that when the Tb^{3+} concentration varies in the range of 0.4–0.8 mol%, the green light intensity of upconversion increases with the increase of Tb^{3+} concentration, and the energy transfer probability from Yb^{3+} to Ho^{3+} decreases, so the red light intensity of upconversion (661 nm) decreases. When the Tb^{3+} concentration increases to

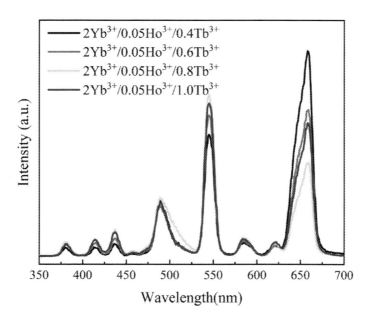

Fig. 6. Upconversion luminescence spectra of phosphate glass samples #5, #6, #7, and #8 under 980 nm laser excitation.

1.0 mol%, the green light intensity decreases, mainly because the gradual formation of Tb^{3+} ion clusters inhibits the energy transfer efficiency between Yb^{3+}-Tb^{3+} ions with the increasing Tb^{3+} doping concentration. Furthermore, the probability of cross-relaxation Tb^{3+} ions increases with the following process: $^5D_3 + {}^7F_6 \rightarrow {}^5D_4 + {}^7F_0$, which leads to a decrease in the number of particles filling the $^5D_3$ energy level and an increase in the number of particles in the $^5D_4$ energy level, and since the $^5D_4$ of Tb^{3+} is similar to the $^5S_2$, $^5F_5$ energy level difference of Ho^{3+}, part of the energy will be transferred from the $^5D_4$ energy level to the $^5S_2$, $^5F_5$ energy level, resulting in the red light intensity (661 nm) of Ho^{3+} being enhanced instead.

By further analysis of the effect of Ho^{3+} concentration on the upconversion luminescence performance of $Yb^{3+}/Ho^{3+}/Tb^{3+}$ tri-doped phosphate glasses, we tested the fluorescence decay lifetime at 545 nm (Tb^{3+}: $^5D_4 \rightarrow {}^7F_5$) for different doping concentrations of Ho^{3+} under 980 nm excitation, and the experimental results are shown in Fig. 7. The relevant calculation formula is as follows[22]:

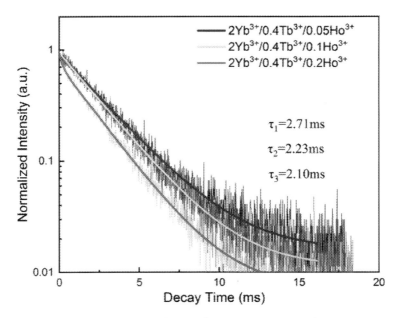

Fig. 7. Decay dynamics of the $^5D_4 \rightarrow {}^7F_5$ transition of Tb^{3+} ions in the glass samples #2, #3, and #5.

$$I(t) = I(0) + A_1 \exp(-t/\tau_1) + A_2 \exp(-t/\tau_2). \tag{1}$$

The lifetime of upconverted green light (545 nm) decreases from 2.71 ms to 2.10 ms as the concentration of Ho^{3+} increases, indicating that the luminescence intensity of Tb^{3+} can be effectively reduced by adjusting the concentration of Ho^{3+}.

For demonstration of the effect of Tb^{3+} ion concentration on the luminescence lifetime of Tb^{3+} ion upconversion, we also tested the fluorescence decay lifetime of Ho^{3+} at 661 nm ($^5F_5 \rightarrow {}^5I_8$) under 980 nm excitation, and the results are shown in Fig. 8. The upconversion luminescence lifetime of Ho^{3+} ions decreases from 0.47 ms to 0.38 ms as the Tb^{3+} ion concentration increases, indicating that the Tb^{3+} concentration can affect the energy transfer efficiency between Yb^{3+}-Ho^{3+}, thereby reducing the Ho^{3+} luminescence intensity.

The variation in the upconversion luminescence intensity as a function of excitation power for the $2Yb^{3+}/0.05Ho^{3+}/0.8Tb^{3+}$

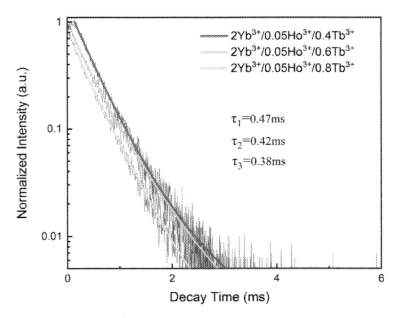

Fig. 8. Decay dynamics of the $^5F_5 \rightarrow {}^5I_8$ transition of Ho^{3+} ions in the samples #5, #6, and #7.

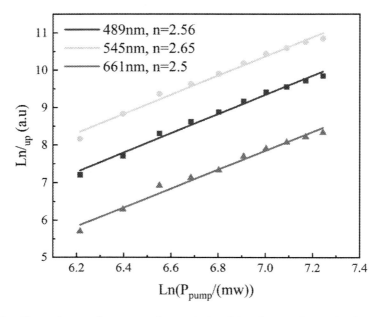

Fig. 9. Dependence of upconversion emission intensity on the excitation power for $Yb^{3+}/Ho^{3+}/Tb^{3+}$ tri-doped phosphate glass.

tri-doped phosphate glass sample is shown in Fig. 9. The curves are fitted with the following equation:

$$I \propto P^n, \qquad (2)$$

where I is the upconversion luminescence intensity, P is the excitation power, and n is the number of photons required[23] for emitting blue (489 nm) and green (545 nm) light by the Tb^{3+} ions and red (661 nm) light by the Ho^{3+} ions. The actual data in the figure have been fitted, and the calculated slope values (n) are 2.56, 2.65, and 2.5, respectively, indicating that the blue and green light emissions are primarily three-photon processes, while the red light emission is a two-photon process.

In order to elucidate the luminescence process of rare-earth tri-doped phosphate glasses, the analysis of the inter-ion energy transfer method in combination with the energy level structural diagram of $Yb^{3+}/Ho^{3+}/Tb^{3+}$ is presented in Fig. 10.

Preparation and upconversion luminescence properties 27

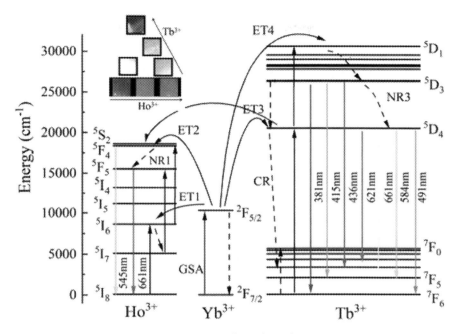

Fig. 10. Energy level diagrams for $Yb^{3+}/Ho^{3+}/Tb^{3+}$ tri-doped phosphate glass under 980 nm excitation.

Firstly, the luminescence mechanism between Yb^{3+}-Ho^{3+} ions is discussed. Under 980 laser excitation, Yb^{3+} ions absorb the pump energy and transition from the ground state $^2F_{7/2}$ to the EXCITED state $^2F_{5/2}$, and the Yb^{3+} ions located in the excited state transfer energy to Ho^{3+}, so that the Ho^{3+} ions completely populate the $^5I_6$ energy level, and the specific process is as follows:

$$^2F_{5/2}(Yb^{3+}) + {}^5I_8(Ho^{3+}) \rightarrow {}^2F_{7/2}(Yb^{3+}) + {}^5I_6(Ho^{3+}).$$

There are two ways for Ho^{3+} ions to realize green light emission. The first one is the absorption of Ho^{3+} ions in the excited state ($^5I_6$ + a 980 nm photon $\rightarrow$ $^5F_4/^5S_2$), and the second one includes the energy transfer between Yb^{3+} and Ho^{3+} ions, where Yb^{3+} transfers the energy of the absorbed photon to the intermediate energy level $^5I_6$ of Ho^{3+}, causing the ion to transition to a higher energy level ($^5F_4/^5S_2$) through the following process:

$$^2F_{5/2}(Yb^{3+}) + {}^5I_6(Ho^{3+}) \rightarrow {}^2F_{7/2}(Yb^{3+}) + {}^5F_4/{}^5S_2(Ho^{3+}).$$

These two processes can populate the $^5F_4/{}^5S_2$ energy level of Ho^{3+}, and finally Ho^{3+} ions located at the excited state $^5F_4/{}^5S_2$ relax to the ground state $^5I_8$, emitting green light at 545 nm.

For achieving red light (661 nm) emission, there are two upconversion mechanisms. Firstly, the particle located in the excited state $^5I_6$ transitions to the lower energy level $^5I_7$ through non-radiative relaxation and then absorbs energy from the Yb^{3+} ions to transition from the energy level $^5I_7$ to $^5F_5$. Another possibility is that the particle in the excited state $^5I_6$ continues to absorb energy from Yb^{3+} ions and then transitions to higher energy levels $^5F_4/{}^5S_2$, Subsequently, the particles in the $^5F_4/{}^5S_2$ energy level populate the $^5F_5$ energy level through non-radiative relaxation, and finally the particles in the excited state $^5F_5$ relax to the ground state $^5I_8$ through radiative transition, emitting red light.

Now, the mechanism of Yb^{3+}-Tb^{3+} inter-ion luminescence is discussed. Under the excitation of 980 NIR light, Yb^{3+} ions transition from the ground state $^2F_{7/2}$ to the excited state $^2F_{5/2}$ by absorbing pump photons through the following process:

$$^2F_{7/2}(Yb^{3+}) + \text{a 980 nm photon} \rightarrow {}^2F_{5/2}(Yb^{3+}).$$

The Yb^{3+} ions in the excited state continue to absorb the pump photons and cause Tb^{3+} ions to transition to the $^5D_4$ energy level by energy transfer. This process is expressed as follows:

$$^2F_{5/2}(Yb^{3+}) + \text{a photon} + {}^7F_6(Tb^{3+}) \rightarrow {}^2F_{7/2}(Yb^{3+}) + {}^5D_4(Tb^{3+}).$$

Meanwhile, the two Yb^{3+} ions located in the excited state $^2F_{5/2}$ can transfer energy to the adjacent Tb^{3+} ions by cooperative upconversion, so that the Tb^{3+} ions excite to a higher energy level and populate the excited state $^5D_4$ ($2Yb^{3+}$: $^2F_{5/2} \rightarrow {}^2F_{7/2}$, Tb^{3+}: $^7F_6 \rightarrow {}^5D_4$). The radiative transition of Tb^{3+} ions from the $^5D_4$ energy level to the lower energy level ^7F_J ($J = 6, 5, 4, 3$) is then completed with the emission of blue, green, yellow, and red light, respectively. It is observed that some of the Tb^{3+} ions at the $^5D_4$ energy level further absorb the energy from the Yb^{3+} ions and continue to transition to the higher energy level $^5D_1$. This process is expressed as follows:

$$^2F_{5/2}(Yb^{3+}) + {}^5D_4(Tb^{3+}) \rightarrow {}^2F_{7/2}(Yb^{3+}) + {}^5D_1(Tb^{3+}).$$

In addition, the Yb^{3+} ions in the ground state can be excited to a higher energy level by directly absorbing three pump photons, causing the transition of Tb^{3+} ions to the $^5D_1$ energy level through energy transfer ($3Yb^{3+}$: $^2F_{5/2} \rightarrow {}^2F_{7/2}$; Tb^{3+}: $^7F_6 \rightarrow {}^5D_1$). Finally, the Tb^{3+} ions on the $^5D_1$ energy level transition to the $^5D_3$ energy level through multi-photon relaxation and then reach the lower energy level ^7F_J ($J = 6$, 5, 4) to emit light with different wavelengths in the range of 350–450 nm.

Next, the quantum efficiency of $2Yb^{3+}/0.05Ho^{3+}/0.6Tb^{3+}$ tri-doped phosphate glass is tested using the following equation:

$$\eta(\lambda) = \frac{Nh\upsilon}{\Phi(\lambda)}. \qquad (3)$$

The results of the test are shown in Fig. 11, $\eta_{(\lambda\,=\,545\text{ nm})} = 36.8\%$, $\eta_{(\lambda\,=\,661\text{ nm})} = 28\%$.

Finally, we discuss the effect of rare-earth ion concentration on luminescence color, the CIE1931 chromaticity coordinates for the upconversion emission of $Yb^{3+}/Ho^{3+}/Tb^{3+}$ tri-doped phosphate glasses under 980 nm excitation are calculated using the following equation:

$$x = \frac{X}{X+Y+Z}, \quad y = \frac{Y}{X+Y+Z}, \quad z = \frac{Z}{X+Y+Z}, \qquad (4)$$

where X, Y, and Z are the three tristimulus values, and the tristimulus values of a color with the spectral power distribution $P(\lambda)$ are given by

$$X = \int_{400}^{720} P(\lambda)x'(\lambda)d\lambda,$$

$$Y = \int_{400}^{720} P(\lambda)y'(\lambda)d\lambda, \qquad (5)$$

$$Z = \int_{400}^{720} P(\lambda)z'(\lambda)d\lambda.$$

30   T. Zhang et al.

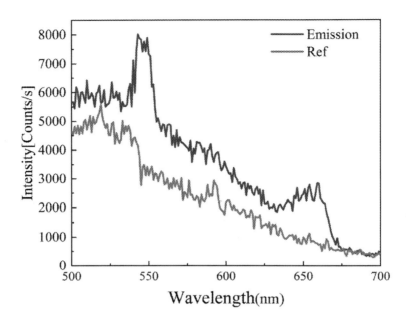

Fig. 11. The quantum efficiency of $2Yb^{3+}/0.05Ho^{3+}/0.6Tb^{3+}$ tri-doped phosphate glass under 980 nm excitation.

Here, λ is the wavelength of the equivalent monochromatic light, and $x'(\lambda)$, $y'(\lambda)$, $z'(\lambda)$ are the three color matching functions.[24]

In Fig. 12(a), the color coordinates (x, y) for the samples #2, #3, and #5 are (0.39, 0.39), (0.52, 0.38), and (0.35, 0.43), respectively. It can be seen that the color coordinates shift from the yellow region to the red region as the Ho^{3+} concentration increases. This result indicates that the luminescence color can be effectively adjusted by varying the Ho^{3+} ion concentration, where the doping concentration that can best match the white light emission is $2Yb^{3+}/0.1Ho^{3+}/0.4Tb^{3+}$. Furthermore, it can be seen in Fig. 12(b) that the color coordinates shift from the yellow region to the green region with the increase in the Tb^{3+} concentration under 980 nm excitation, and the chromaticity coordinates of the four samples (#5, #6, #7, and #8) are (0.35, 0.43), (0.35, 0.47), (0.30, 0.48), and (0.34, 0.46), respectively. This also verifies that phosphate

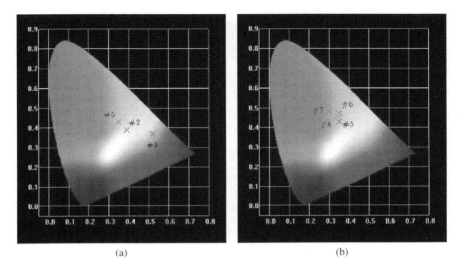

Fig. 12. Calculation of the color coordinates of upconversion emission under 980 nm excitation for the $Yb^{3+}/Ho^{3+}/Tb^{3+}$ tri-doped phosphate glasses. (a) Effect of xHo^{3+} (x = 0.05–0.2 mol%) concentration on the luminescence color; (b) Effect of yTb^{3+} (y = 0.4–1.0 mol%) concentration on the luminescence color.

glasses can emit different colors of light by doping them with different concentrations of Tb^{3+}.

4. Conclusions

$Yb^{3+}/Ho^{3+}/Tb^{3+}$ tri-doped phosphate glasses were prepared by high-temperature melting method using P_2O_5 as the base material, and the prepared glasses were characterized by X-ray diffraction. The dependence of upconversion emission intensity on the excitation power was tested to confirm that the emission of blue and green light by Tb^{3+} was a three-photon process. Furthermore, the experimental results provided useful insights on achieving the target colors. Particularly, it was observed that the CIE chromaticity coordinates could be modulated by adjusting the doping concentration of rare-earth ions, which verified that the $Yb^{3+}/Ho^{3+}/Tb^{3+}$ tri-doped La_2O_3-P_2O_5 phosphate glass exhibits immense application prospect in the illumination field, including display devices, fiber lasers, etc.

Acknowledgements

Fund Projects: This work was supported by Innovation project of GUET Graduate Education No. 2020YX(16), partially supported by the Fund of Guangxi Key Laboratory or Wireless Broadband Communication and Signal processing (No. PF19113P) and by the Foundation of Guilin University of Electronic Technology (No. HKY19002).

References

1. J. F. Suyver, A. Aebischer, D. Biner, P. Gerner, J. Grimm, S. Heer, K. W. Krämer, C. Reinhard and H. U. Güdel, *Opt. Mater.* **27**, 1111 (2005).
2. B. S. Richards, *Sol. Energ. Mat. Sol. C.* **90**, 1189 (2006).
3. C. Sanchez, B. Lebeau, F. Chaput and J. P. Boilot, *Adv. Mater.* **15**, 1969 (2003).
4. N. Bloembergen, *Phys. Rev. Lett.* **2**, 84 (1959).
5. P. Wang, P. Joshi, A. Alazemi and P. Zhang, *Biosens. Bioelectron.* **62**, 120 (2014).
6. F. Wang, Y. Han, C. S. Lim, Y. Lu, J. Wang, J. Xu, H. Chen, C. Zhang, M. Hong and X. Liu, *Nature* **463**, 1061 (2010).
7. Q. Nie, Z. Jin, T. Xu, S. X. Dai, X. Shen and X. H. Zhang, *BullChin. Ceram. Soc. (in Chinese)* **5**, 108 (2006).
8. M. Yousaf, M. N. Akhtar, B. Wang and A. Noor, *J. Tec. Sci.* **46**, 4280 (2020).
9. Z. Wang, F. Huang, B. Li, Y. Li, Y. Tian and S. Xu, *News Sci.* **219**, 116949 (2020).
10. L. Debasu Mengistie, A. Duarte, R. João, L. Malta Oscar and D. Carlos Luís, *PCCP* **15**, 15565 (2013).
11. A. El hat, I. Chaki, R. Essajai, A. Mzerd, G. Schmerber, M. Regragui, A. Belayachi, Z. Sekkat, A. Dinia, A. Slaoui and M. Abd-Lefdil, *Crystals* **10**, 169 (2020).

12. J. Liu, Q. Song, J. Xu, J. Guo, Y. Pan, N. Li, D. Li, P. Liu, X. Xu, J. Xu and K. Lebbou, *J. Lumin.* **246**, 118826 (2022).
13. Y. Chu, Y. Yang, Z. Liu, L. Liao, Y. Wang, J. Li, H. Li, J. Peng, N. Dai, J. Li and L. Yang, *Appl. Phy. A.* **118**, 1429 (2015).
14. R. Zhang, H. Lin, Y. Yu, D. Chen, J. Xu and Y. Wang, *Laser. Photonics. Rev.* **8**, 158 (2014).
15. J. Wang, C. Tsai, W. C. Cheng, M. H. Chen, C. H. Chung and W. H. Cheng , *Ieee. J. Sel. Top. Quant.* **17**, 741 (2011).
16. S. Long, D. Ma, Y. Zhu, M. Yang and S. Li, *J. Lumin.* **192**, 728 (2017).
17. Z. Lin, H. Lin, J. Xu, F. Huang, H. Chen, B. Wang and Y. Wang, *J. Eur. Ceram. Soc.* **36**, 1723 (2016).
18. G. Lakshminarayana, H. Yang and J. Qiu, *J. Solid. State. Chem.* **182**, 669 (2009).
19. C. Zhu, X. Liang, Y. Yang and G. Chen, *J. Lumin.* **130**, 74 (2010).
20. D. Rajesh, K. Brahmachary, Y. C. Ratnakaram, N. Kiran, A. P. Baker and G. G. Wang, *J. Alloy. Compd.* **646**, 1096 (2015).
21. L. Ren, X. Lei, X. Du, L. Jin, W. Chen and Y. Feng, *J. Lumin.* **142**, 150 (2013).
22. V. Pankratov, A. I. Popov, S. A. Chernov, A. Zharkouskaya and C. Feldmann, *Phys. Status. Solidi. B* **247**, 2252 (2010).
23. H. Guo, N. Dong, M. Yin, W. Zhang, L. Lou and S. Xia, *J. Phys. Chem. B.* **108**, 19205 (2004).
24. H. Gong, D. L. Yang, X. Zhao, E. Y. B. Pun and H. Lin, *Opt. Mater.* **32**, 554 (2010).

Chapter 3

PVA-assisted synthesis of Zn_2SiO_4: Mn^{2+} nanoparticles: Enhanced luminescence properties and applications

Huiya Li*, Yaoyao Li*, Xiaomeng Wang*, Zicheng Zhang*, Dejia Liu*,
Haiyun Ma[†,§,**], Hongqiang Qu[†], Jianzhong Xu[†],
Yuanyuan Han[‡,¶,**] and Liyong Wang*[‖,**]

*College of Chemistry and Materials Science
Hebei University, Baoding 071002, P. R. China
[†]College of Chemistry and Material Science Flame Retardant Material and
Processing Technology Engineering Research Center
Baoding 071002, P. R. China
[‡]Medical Experimental Centre of Hebei University
Baoding, 071000, P. R. China
[§]coffee1123@126.com
[¶]hanyy_8016@163.com
[‖]wangly_1@126.com; wangly@hbu.cn

Green-emitting Mn^{2+}-doped zinc silicate nanophosphor was synthesized by solvothermal method with the assistance of polyvinyl alcohol (PVA). The structure, morphology and optical properties of phosphors were characterized by X-ray diffraction (XRD), X-ray photoelectron spectroscopy (XPS), Scanning electron microscope (SEM), Fourier transform infrared spectrometer (FT-IR) and Photoluminescence Spectroscopy (PL). Also, the application of Mn^{2+}-doped zinc silicate

---

**Corresponding authors.

To cite this article, please refer to its earlier version published in the Functional Materials Letters, Volume 16(7), 2340022 (2023), DOI: 10.1142/S1793604723400222.

36 *H. Li et al.*

phosphor in electronic devices was studied. Furthermore, Zn_2SiO_4: Mn^{2+} phosphor was used to be a fluorescent sensor for the first time to detect inorganic ions and small molecules. The results indicated that pure α-Zn_2SiO_4 phase nanoparticles with spherical morphology were obtained. The results also showed that the phosphor has excellent selectivity and sensitivity for detecting Cr^{3+} ions, $Cr_2O_7^{2-}$ ions and 4-nitrophenol (4-NP) with low detection limits (Cr^{3+} ions: 1.34 μM, $Cr_2O_7^{2-}$ ions: 5.86 μM, 4-NP: 2.38 μM). Thus, Mn^{2+}-doped zinc silicate manganese phosphor has potential applications in light-emitting diodes (LEDs) and fluorescent sensors.

Keywords: Zinc silicate; polyvinyl alcohol; luminescence; applications.

1. Introduction

In recent years, researchers have been attracted to phosphors because of their ability to provide powerful luminescence in optoelectronic devices such as light-emitting, solid-state lighting and cathode ray tubes.[1 5] However, with the advancement of technology and better competition in the market, phosphors are required to develop toward higher light-emitting efficiency, higher brightness, and smaller size.[6] It is documented that luminescent properties of phosphors are mainly influenced by the host lattice, dopant and synthesis method.[7] In addition, reducing the particle size is also conducive to improving the luminescent properties of luminescent materials. Furthermore, when the particle size is nearly the nanoscale, its chemical and physical properties are superior to micrometer counterparts. To improve the dispersion and luminescent properties of phosphors, it is essential to synthesize nanomaterials with controllable size and morphology.[2,4,8-10]

Zinc silicate (Zn_2SiO_4) is one of the excellent phosphor substrate materials, due to its advantages of wide emission spectra, good chemical stability, high color saturation, high luminescence efficiency, long fluorescence life and strong absorption in the near ultraviolet region.[2,5] Therefore, Zn_2SiO_4 phosphors are widely used in fluorescent lamps, electroluminescent (EL) devices, plasma display panels (PDPs) and cathode ray tubes (CRTs).[2 4,11] It was documented that the energy band of Zn_2SiO_4 is about 5.5 eV, and different colors of Zn_2SiO_4 phosphors can be obtained when doped with different ions.[12] Typically, transition metals and rare earth ions are used as activators to synthesize phosphors in the visible and near-infrared regions.[13] In particular, the divalent transition metal ions enable phosphors to exhibit stable emission due to the d–d electron transition.[5] It is well known that Zn_2SiO_4 is available in α, β and γ crystalline phases. Among them, α-Zn_2SiO_4 is composed of SiO_4^{4-} and ZnO_4^{6-} tetrahedral units, and the doping ions replace zinc in the tetrahedra, thus different emitting phosphors are obtained.[14,15] Zn_2SiO_4:Mn^{2+} is one of the earliest synthesized phosphors.[16] Mn^{2+} ions can be uniformly doped in the zinc silicate matrix for tetra coordination and present green emission, because Mn^{2+} ions and Zn^{2+} ions have the same oxidation state and close radius.[17] Moreover, as an activator, Mn^{2+} ions can reduce the use of rare earth ions, and Zn_2SiO_4:Mn^{2+} is a promising fluorescent material for multi-field applications.[16] There are several synthesis methods for Zn_2SiO_4:Mn^{2+} phosphor, including high-temperature solid-phase,[18] spray pyrolysis,[19,20] sol–gel method,[5,15,21 24] chemical vapor deposition method,[25] ultrasound-assisted method,[4] hydrothermal methods[10,26] and new organic–inorganic polymerization synthesis.[7] However, among these methods, the solvothermal method has the advantages of uniform particle size and controllable morphologies.[27] Nevertheless, compared with the irregular morphology, the spherical morphology possesses higher packing density, lower light scattering degree and well photoluminescence, so the candidate of light-emitting particles should be spherical shape with narrow size distribution.[4,28,29] It is well known that the distinctive functions and characteristics of nanomaterial provide new prospects for technological development and innovation in various practical applications.[13] So far, more effects have been devoted to design new LED phosphors for solid-state lighting. Moreover, Zn_2SiO_4 nano-sized phosphors can also be applied in assays and biomarkers.[30]

Herein, nanoscale spherical Zn_2SiO_4:Mn^{2+} green phosphor was prepared by a low-temperature solvothermal method assisted by polyvinyl alcohol (PVA). The crystal structure, morphology and photoluminescence properties of the phosphor were investigated. Based on the excellent luminescent properties of Zn_2SiO_4:Mn^{2+}, it was applied to LED devices and metal ions assay.

2. Experiment

2.1. *Chemicals*

All chemicals used in the synthesis were of analytical grade without further purification. All the inorganic and organic reagents were purchased from Tianjin Kermel Chemical Reagent Co., Ltd.

2.2. *Synthesis of green-emitting phosphor of $Zn_2SiO_4:Mn^{2+}$*

$(CH_3COO)_2$ $Zn\cdot2H_2O$ and $(CH_3COO)_2Mn\cdot4H_2O$ were added, and 0.200 g of cetyltrimethylammonium bromide (CTAB) and 0.30 mL of pentanol were added in the absence and presence of polyvinyl alcohol (PVA). Next, the well-stirred solutions were transferred to Teflon reaction vessel of the autoclaves, which were subsequently sealed and heated to 160°C for 6 h. After the autoclaves were cooled to room temperature, the products were washed with deionized water and anhydrous ethanol and dried at 100°C. Finally, the products were calcined at 800°C for 6 h to obtain $Zn_2SiO_4:Mn^{2+}$. The $Zn_2SiO_4:Mn^{2+}$ products prepared unassisted and assisted by PVA were noted as ZSM and ZSM-P, respectively.

2.3. *Material characterizations*

X-ray diffraction (XRD) measurement was performed to investigate the phase on a D8 Advanced Bruker X-ray diffractometer with Cu $K\alpha$ radiation (40 KV, 40 MA). The morphologies of samples were measured by a scanning electron microscope (SEM, JSM-7500), and elements of products were measured by using energy dispersive spectrometry (EDS, ce-440). X-ray photoelectron spectroscopy (XPS, Escalab Xi$^+$) was performed to study chemical bonding states and compositions of nanoparticles. FT-IR spectra were measured by using a spectrometer (Nicolet iS10) with the KBr pellet in the region of 4000–400 cm^{-1}. The optical absorption spectra and band gap energy were conducted by using a UV–Vis spectrometer (Shimadzu-UV3600) and collected the spectra in the region of 200–800 nm. Photoluminescence (PL) spectra were measured by using a fluorescence spectrometer (Hitachi F-7000) was used to measure PL spectra. All experimental tests were carried out at room temperature.

2.4. *Detection of inorganic anions, cations and small organic molecules*

To further evaluate the sensing selectivity of Zn_2SiO_4:Mn^{2+} phosphor suspensions for inorganic ions and small organic molecules, 5 mg of ZSM-P phosphor was dispersed in 50 mL of distilled water to prepare suspensions. Then, different inorganic ions and small organic molecules were dissolved in deionized water to prepare solutions. Next, 100 μL of pre-prepared solutions were added to 1500 μL of the suspension and PL spectra were obtained at 255 nm. To study the sensitivity of ZSM-P suspension, 100 μL of different concentrations of ions and small molecule solutions were added to 1500 μL of suspension and the fluorescence intensities were recorded at 255 nm.

2.5. *Assembly of LED optics*

The appropriate amount of ZSM and ZSM-P powders were transferred to the 280 nm LED chip slot for compaction. Afterwards, the positive and negative terminals of the chip were connected to assemble the LED devices and measured at 5.6V.

3. Results and Discussions

Figure S1(a) shows XRD patterns of ZSM and ZSM-P and the results show that both of them process similar diffraction patterns. Furthermore, the peaks are well indexed to the standard structure of Zn_2SiO_4 (JCPDS No.70–1235) and no impurity peak appears in the patterns. The crystal systems belong to the rhombohedral system with the space group R3 (No. 148) and cell parameters are a (13.948 Å), b (13.948 Å), c (9.315 Å), α (90°), β (90°) and γ (120°), respectively.

40   H. Li et al.

From these results, it can be concluded that willemite was successfully prepared. Also, since the ionic radii are very close ($r\text{Mn}^{2+}$ = 0.83 Å and $r\text{Zn}^{2+}$ = 0.74 Å) and no other phases are detected, indicating that Mn^{2+} is uniformly mixed and effectively doped at the Zn^{2+} ions sites in the matrix lattice.[31] Figure S1(b) shows the XRD patterns of with different Mn ion doping concentrations. With the increase of Mn ion concentrations, the intensities of diffraction peaks first enhanced and then decreased. The strongest diffraction peaks are found when the doping concentration is 7%. In addition, XRD Rietveld refinements of ZSM and ZSM-P were implemented in order to obtain further crystal structural information. The XRD refinement plots of ZSM and ZSM-P, depicted in Figs. 1(a) and 1(b) were performed by GSAS-EXPGUI software.[32]

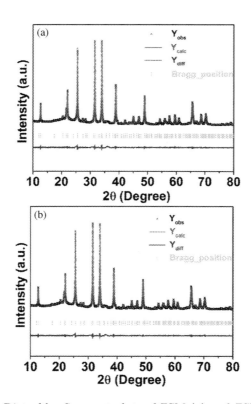

Fig. 1.   Rietveld refinement plots of ZSM (a) and ZSM-P (b).

Table 1. Refinement results and structural parameters of ZSM and ZSM-P.

Parameter	ZSM	ZSM-P
Cell parameters		
a (Å)	13.9378(1)	13.9388(1)
b (Å)	13.9378(1)	13.9388(1)
c (Å)	9.3104(1)	9.3110(1)
(α, β, γ)	(90°,90°,120°)	(90°,90°,120°)
V (Å^3)	1566.34(3)	1566.68(4)
Crystal system	Rhombohedral	Rhombohedral
Space group	$R3$	$R3$
Refinement parameters		
Rwp (%)	4.78	4.62
Rp (%)	3.60	3.52
χ^2	2.011	1.877

The refined cell parameters, reliability factors and other relevant information are demonstrated in Table 1. The refined atomic coordinate parameters of ZSM and ZSM-P are shown in Tables S1 and S2. The cell parameters of ZSM and ZSM-P showed a slight variation, and it can be attributed to the stress and strain generated by PVA on the nanoparticle surface.[33] The obtained reliability factors Rwp and Rp are well below 10%, thus indicating that the results after refinement are reliable.

XPS was used to study the binding states and the compositional elements of ZSM and ZSM-P. As shown in Fig. 2(a), XPS peaks of O1s, Si_2p, Zn_2p and Mn_2p can be observed in the full-scanning spectra of ZSM and ZSM-P, confirming successful synthesis of Zn_2SiO_4:Mn^{2+}. As shown in Fig. 2(b), the O1s peaks of ZSM and ZSM-P are both presented with two fitted peaks. The fitted peak at 532.7 eV is attributed to the Si–O–Si bond, and the other fitted peak at 531.3 eV is related to the O–Zn bond.[34] As seen from

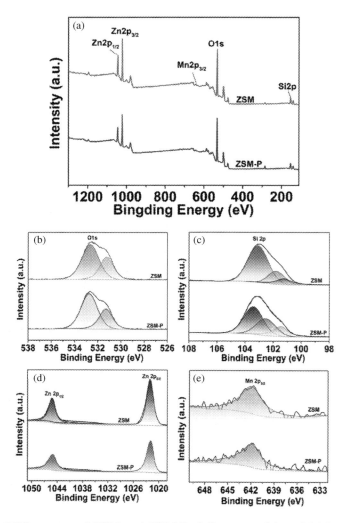

Fig. 2. XPS spectra of ZSM and ZSM-P: full spectra (a) and high-resolution spectra of Si2p (b), O1s (c), Zn2p (d) and Mn2p (e).

Fig. 2(c), the Si$_2$p peaks of ZSM are broad and asymmetric, which could be fitted into three peaks at 103.1 eV (SiO$_4$), 101.8 eV (Si–O–Si) and 101.25 eV (Si–OH).[29] Correspondingly, the binding energy values of Si 2p in ZSM-P sample are 103.4, 102.55 and 101.45 eV. As shown in Fig. 2(d), the Zn$_2$p orbital signal exhibit two characteristic peaks. The $Zn_2p_{1/2}$ and $Zn_2p_{3/2}$ peaks of ZSM and ZSM-P are observed at 1045eV and 1022 eV, respectively. The former and latter peaks could be ascribed to the fact that Zn^{2+} ions, respectively, occupy the tetrahedral and octahedral sites.[35] Seen from Fig. 2(e), the peaks at 642.05 (ZSM) and 641.9 (ZSM-P) eV are assigned to the orbital of $Mn2p_{3/2}$. In addition, the XPS spectra of Mn_2p show a high signal-to-noise ratio due to low concentration of Mn doping in the main lattice.[35]

The SEM images, corresponding elemental mapping images and EDS spectra of ZSM and ZSM-P are presented in Fig. 3. It can be observed that both exhibit spherical shapes of smaller than 100 nm (seen from Figs. 3(a) and 3(b)). Moreover, the addition of PVA has no significant effect on their morphology. To further investigate the effect of PVA on particle size, dynamic light scattering (DLS) was performed. The average particle diameters of ZSM and ZSM-P are detected to be 342 nm and 220 nm, respectively (seen from Fig. S2). Although the phenomenon of agglomeration appears, it can be concluded that the particle size of $Zn_2SiO_4:Mn^{2+}$ has been reduced with

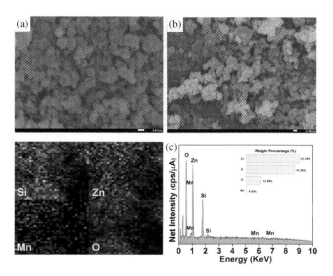

Fig. 3. SEM micrographs of ZSM (a) and ZSM-P (b). Elemental mapping and EDS spectra of ZSM-P (c).

the assistance of PVA. In addition, the composition and distribution of each element are presented in Fig. 3(c). The elemental mapping clearly shows that Zn, Si, O, and Mn are dispersed throughout the region, further confirming the successful synthesis of Zn_2SiO_4: Mn^{2+}, which is consistent with the XRD analysis.

Figure S3 presents FT-IR spectra of ZSM and ZSM-P. Basically, both spectra show the same pattern, indicating that PVA in ZSM-P has been removed thoroughly after calcination. The peak observed at 3426 cm^{-1} is related to the stretching vibration of O–H group.[16] The absorption peak at 1100 cm^{-1} is attributed to the Si–O–Si asymmetric stretching vibrations mode.[36] The peaks are observed at 981, 929 and 900 cm^{-1} due to the asymmetric stretching vibration of SiO_4^{4-}.[5,23] The vibrational band appeared at 867 cm^{-1} is ascribed to the symmetric stretching vibration of SiO_4^{4-}.[23,37] The bands at 619 cm^{-1} and 576 cm^{-1} are induced by asymmetric and symmetric tensile vibrations of ZnO_4^{6-}.[23,36] The band at 461 cm^{-1} corresponds to asymmetric deformation of SiO_4^{4-}.[5,23] It is clear that a willemite (Zn_2SiO_4) structure is ascribed to the existence of characteristic vibration of SiO_4^{4-} and ZnO_4^{6-} groups, which is consistent with XRD and XPS results.

To evaluate the band gap of ZSM and ZSM-P, UV–Vis diffuse reflectance analysis was performed and the spectra were demonstrated in Fig. S4. As shown in Fig. S4(a), the absorption peaks appeared at 238 nm can be attributed to the energy transfer from Zn^{2+} to Mn^{2+} in short wavelength.[7] The energy band (E_g) can be obtained by the Kubelka–Munk or the Tauc plot method.[12,21,24,38,39] The energy gap (E_g) can be calculated according to the following equation by the Tauc plot method:

$$(\alpha h\nu)^{1/n} = B(h\nu - E_g), \tag{1}$$

where α is an absorption coefficient, $h\nu$ denotes photon energy, n is related to the semiconductor type, B is an absorption constant.[1] It is documented that Zn_2SiO_4 is an indirect band gap material with the value of n is $1/2$.[39] Figure S4(b) reflects the relationship between $(\alpha h\nu)^2$ and $h\nu$. Energy gap values of ZSM and ZSM-P are 4.50 eV and 4.46 eV, respectively. In general, the band gap decreases with the assistance of PVA. It is assigned to the increased absorption of electrons from the valence band (VB) to the conduction band (CB) because of assisted synthesis of PVA, as well as to the creation of a persistent internal force, thus reducing the optical band gap.[40] Compared to pure Zn_2SiO_4, the band gap of ZSM and ZSM-P decrease. It can be attributed to the exchange interaction between the d electrons of the Mn ions and the s and p electrons of the host band.[21]

Figure 4(a) depicts the excitation and emission spectra of Zn_2SiO_4: Mn^{2+}. As shown in Fig. 4(a), the excitation spectrum shows two broad absorption bands. The band from 200 nm to 225 nm

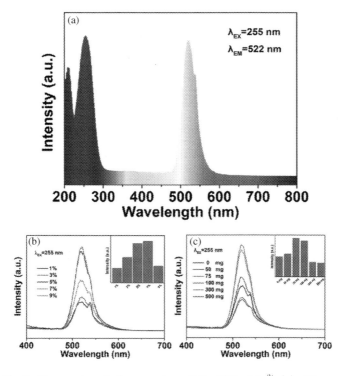

Fig. 4. Excitation and emission spectra of Zn_2SiO_4: Mn^{2+} (a). PL spectra and histograms of Zn_2SiO_4: Mn^{2+} as a function of Mn^{2+} doping (b). PL spectra and histograms of Zn_2SiO_4: Mn^{2+} with different amounts of PVA (c).

is ascribed to the energy transfer from the host lattice (Zn_2SiO_4) to the activator Mn^{2+} ions.[3] The other absorption peak excited at 255 nm is attributed to the energy transfer from the ground state $^6A_1$ (^6S) to the conduction band (CB) of the luminescent center Mn^{2+} ions.[24,41] Figure 4(a) exhibits the strongest broad emission at 522 nm, ascribing to the spin-forbidden $^4T_1-^6A_1$ leap of the $3d^5$ electron of the Mn^{2+}.[10,15,25] In addition, there is a shoulder at 538 nm in the PL emission spectra. It can be explained by the radiative recombination from the donor (interstitial Zn defect and O vacancy) to the acceptor (cationic Zn vacancy) and the $^4T_1(G) \rightarrow ^6A_1(S)$ transition of Mn^{2+}.[12]

To obtain the optimal Zn_2SiO_4:Mn^{2+} phosphor, the optimal reaction conditions for the synthesis were explored. The doping concentration of Mn^{2+} ions is one of the important influencing factors on fluorescence intensity. Figure 6(b) shows the variation of the emission intensities with the doping concentration of Mn^{2+} ions. When the doping concentrations of Mn^{2+} increase from 1% to 9%, the positions of the emission peaks have no shift. However, the fluorescence intensities first increase and then decrease. ZSM-P had the strongest green emitting when the doping concentration of Mn^{2+} was 7%. The optimal doping concentration of Mn^{2+} is determined by the competition between increased luminescence centers and lower luminescence efficiency.[3,20] Moreover, the amount of PVA is an influential factor for the luminescence performance of the samples. Figure 4(c) shows the emission intensities of Zn_2SiO_4: Mn^{2+} with different amounts of PVA. It can be observed that the emission intensities of Zn_2SiO_4:Mn^{2+} enhance first and then reduces with the increase of PVA amount, and the emission intensity reaches the maximum at the amount of 0.075 g of PVA. The significant optimization of the fluorescence performance is attributed to the addition of PVA surfactant, which reduces the surface defects and the reflectance loss at the particle-air interface of the samples.[42]

To investigate the application of ZSM-P, a series of cations, anions and organic molecules were used to probe the response to fluorescence intensity. Figure 5 illustrates the emission spectra and fluorescence intensities of ZSM-P suspensions (λ_{ex} = 255 nm) with

PVA-assisted synthesis of Zn_2SiO_4: Mn^{2+} nanoparticles 47

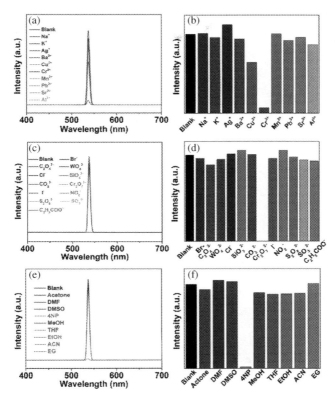

Fig. 5. Emission spectra and intensities of ZSM-P suspensions (λ_{ex} = 255 nm) with the addition of different cations (a), (b), anions (c), (d) and organic reagents (e), (f).

the addition of different cations, anions and organic molecules. As shown in Fig. 5, most of them show slight fluorescence quenching for ZSM-P except for Cr^{3+}, $Cr_2O_7^{2-}$, and 4-nitrophenol (4-NP). The results indicate that ZSM-P can be a potential candidate for Cr^{3+}, $Cr_2O_7^{2-}$ and 4-NP assay. To investigate luminescence quenching mechanism, UV–Vis absorption spectra of Cr^{3+}, $Cr_2O_7^{2-}$ and 4-NP were recorded. It can be observed from Fig. S5(a) the overlaps between the excitation and emission spectra of ZSM-P and the UVVis absorption spectrum of Cr^{3+}. It is a competition between the absorption of Cr^{3+} ions and the excitation and emission of ZSM-P, resulting in luminescence quenching.[43] Seen from Fig. S5(b), the absorption peak of $Cr_2O_7^{2-}$ can be observed at 256 nm, and the absorption spectrum overlaps with the excitation spectrum (λ_{ex} = 255 nm) of ZSM-P. The quenching mechanism can be attributed to the competitive absorption of excitation wavelength energy between ZSM-P and $Cr_2O_7^{2-}$ [44,45] as shown in Fig. S5(c), the excitation of ZSM-P overlap with the absorption peak of 4-NP, and the quenching effect is caused by the competitive absorption of 4-NP, which reduces the energy transfer efficiency.[46]

It can be seen from Figs. 6(a), 6(c) and 6(e) that the fluorescence intensities are gradually quenched by Cr^{3+}, $Cr_2O_7^{2-}$ and 4-NP with the concentrations of targets increase. To quantify the sensitivity of ZSM phosphor for Cr^{3+}, $Cr_2O_7^{2-}$ and 4-NP assay, the

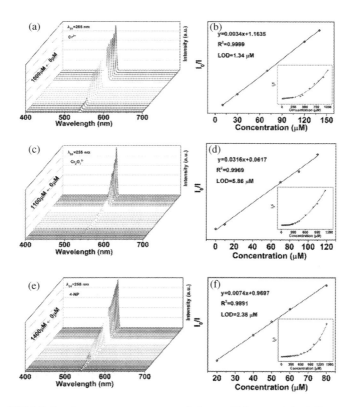

Fig. 6. ZSM-P luminescence spectra and Stern–Volmer with different concentrations of Cr^{3+} (a), (b), $Cr_2O_7^{2-}$ (c), (d) and 4-NP (e), (f).

fluorescence quenching process can be described by the Stern–Volmer equation. The Stern–Volmer equation is displayed in the following equation:

$$I_0/I = 1 + K_{sv} \ C, \tag{2}$$

and targets suspension, C is the concentration of targets, and K_{sv} is Stern–Volmer constant.[47] As shown in Fig. 6(b), the value of I_0/I increases with increasing concentration of Cr^{3+} ions in 0–1000 μM range. When the concentration is in the range of 0–150 μM, both exhibit a good linear relationship with a Ksv value of 3.4 × 10^{-3} (R^2 = 0.9999). In addition, the limit of detection (LOD = $3\sigma/K_{sv}$,[47] σ is the standard deviation of the slope, Ksv is the absolute value of the slope) can be calculated from the absolute value of the slope. In Fig. 6(b), the LOD of Cr^{3+} is calculated to be 1.34 μM. Similarly, seen from Fig. 6(d), the value of I_0/I keeps increasing with the increase of $Cr_2O_7^{2-}$ ions in the range of 0–1100 μM. The LOD of $Cr_2O_7^{2-}$ ions is 5.86 μM with a Ksv value of 0.0316 (R^2 = 0.9969) in the range of 0–110 μM. Figure 6(f) shows that the value of I_0/I increases with increasing concentration of 4-NP in the range of 0–1400 μM. When the concentration is in the range of 0–80 μM, both of them show a good linearity with a Ksv value of 7.3 × 10^{-3} (R^2 = 0.9991) and LOD of 2.38 μM for 4-NP. It can be found that ZSM-P as a fluorescent sensor show a wide detection range and low detection limit. Specifically, comparison of ZSM-P with other fluorescent sensors toward Cr^{3+}, $Cr_2O_7^{2-}$ and 4-NP is provided in Table S3 and the results are better than those previously published papers. All results reveal that ZSM-P can be acted as a multi-responsive sensor for Cr^{3+}, $Cr_2O_7^{2-}$ and 4-NP with high selectivity and sensitivity.

To expand the application of materials, LED optical devices were fabricated by ZSM and ZSM-P powder. LED device photographs are shown in Fig. 7. It can be seen from Figs. 7(b) and 7(d) that both of LED devices emit green light at 5.6 V. Furthermore, compared with assembled LED device by ZSM, LED device composed of ZSM-P emit brighter and stronger green light. The results

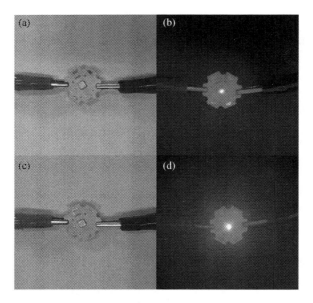

Fig. 7. LED devices prepared with ZSM: in unenergized condition (a) and energized (b). LED devices equipped with ZSM-P: in unenergized (c) and energized state (d).

show that PVA-modified synthesis of Zn_2SiO_4: Mn^{2+} phosphor significantly improves luminescence performance. Thus, ZSM-P could be used as an efficient green luminescent material, indicating potential applications in light-emitting diodes.

4. Conclusion

In conclusion, green-emitting Zn_2SiO_4:Mn^{2+} was synthesized by a low-temperature solvothermal method with the assistance of polyvinyl alcohol (PVA). The comprehensive characterization of the structure, morphology and PL properties showed that nanoscale Zn_2SiO_4:Mn^{2+} with spherical shape was successfully prepared. In addition, PVA adding enhanced remarkably the luminescence of Zn_2SiO_4:Mn^{2+} phosphor. The optimal photoluminescence of Zn_2SiO_4:Mn^{2+} was obtained with 0.075 g of PVA and 7% of Mn2+ (λ_{ex} = 255 nm, λ_{em} = 522 nm). Based on excellent light-emitting properties, the as-prepared has potential applications for LED devices. Furthermore, $Zn_2SiO_4:Mn^{2+}$ is selective and sensitive for Cr^{3+}, $Cr_2O_7^{2-}$ and 4-NP assay, which can be as a promising material for detection.

Acknowledgments

This work was supported by National Natural Science Foundation of China (Grant No. 52173073), Students Research Fund and Graduate project Innovation of Hebei University 2020273, sy202001, 2021253, 2022225, Medical Science Foundation of Hebei University 2021B12, Multi-disciplinary Research Project of Hebei University DXK202003, Beijing Tianjin Hebei Collaborative Innovation Community Construction Project 20541401D, Key project of Natural Science Foundation of Hebei Province E2021201003.

References

1. S. H. Jaafar *et al.*, *Crystals* **11**, 115 (2021).
2. C. W. Mun *et al.*, *J. Mol. Struct.* **1248**, 131474 (2022).
3. J. B. Yoo *et al.*, *J. Lumin.* **243**, 118608 (2022).
4. R. B. Basavaraj *et al.*, *Ultrason. Sonochem.* **34**, 700 (2017).
5. K. Omri *et al.*, *Ceram. Int.* **43**, 6585 (2017).
6. S. Li *et al.*, *Laser Photon. Rev.* **12**, 1800173 (2018).
7. A. Manavbasi and J. C. LaCombe, *J. Mater. Sci.* **42**, 252 (2006).
8. C. E. Rivera-Enríquez *et al.*, *J. Alloys Compd.* **688**, 775 (2016).
9. A. M. Youssef and S. M. El-Sayed, *Carbohydr. Polym.* **193**, 19 (2018).
10. X. Yu and Y. Wang, *J. Alloys Compd.* **497**, 290 (2010).
11. R. M. Krsmanović *et al.*, *Appl. Phys. A* **104**, 483 (2011).
12. X. He *et al.*, *Ceram. Int.* **48**, 19358 (2022).
13. S. M. Abo-Naf and M. A. Marzouk, *Nano-Struct. Nano-Objects* **26**, 100685 (2021).
14. Y. Jiang *et al.*, *Mater. Chem. Phys.* **120**, 313 (2010).
15. P. S. Mbule *et al.*, *J. Lumin.* **179**, 74 (2016).
16. P. Sharma and H. S. Bhatti, *J. Alloys Compd.* **473**, 483 (2009).
17. Q. Zhou *et al.*, *J. Mater. Chem. C* **6**, 2652 (2018).
18. L. M. Xiong *et al.*, *J. Phys. Chem. B* **109**, 731 (2005).

19. S. H. Nam *et al.*, *Funct. Mater. Lett.* **03**, 97 (2011).
20. C. H. Lee *et al.*, *Mater. Sci. Eng. B* **117**, 210 (2005).
21. J. El Ghoul and L. El Mir, *J. Lumin.* **148**, 82 (2014).
22. Z. Ji *et al.*, *J. Cryst. Growth* **255**, 353 (2003).
23. K. Omri *et al.*, *Appl. Phys. A* **124**, 215 (2018).
24. L. El Mir and K. Omri, *Superlattices Microstruct.* **75**, 89 (2014).
25. T. Ohtake *et al.*, *J. Alloys Compd.* **421**, 163 (2006).
26. N. A. Zaitseva *et al.*, *Russ. J. Inorg. Chem.* **62**, 168 (2017).
27. T. J. Lou *et al.*, *J. Colloid Interface Sci.* **314**, 510 (2007).
28. V. Castaing *et al.*, *J. Appl. Phys.* **130**, 080902 (2021).
29. P. S. Mbule *et al.*, *J. Lumin.* **192**, 853 (2017).
30. Y. Li *et al.*, *Talanta* **179**, 420 (2018).
31. H. Liu *et al.*, *Phys. Chem. Chem. Phys.* **20**, 10086 (2018).
32. D. S. Parsons *et al.*, *Inorg. Chem.* **58**, 16313 (2019).
33. T. Dippong *et al.*, *Ceram. Int.* **44**, 7891 (2018).
34. L. Wang *et al.*, *Angew. Chem. Int. Ed. Engl.* **58**, 8103 (2019).
35. D. Q. Trung *et al.*, *J. Alloys Compd.* **845**, 156326 (2020).
36. S. H. Yang *et al.*, *Appl. Phys. A: Mater. Sci. Process* **127**, 588 (2021).
37. K. N. Prathibha *et al.*, *J. Mater. Sci: Mater. Electron.* **32**, 20197 (2021).
38. K. Omri *et al.*, *Physica E* **53**, 48 (2013).
39. Z. Li *et al.*, *Mater. Chem. Phys.* **240**, 122144 (2020).
40. S. A. A. Wahab *et al.*, *J. Alloys Compd.* **926**, 166726 (2022).
41. R. Ye *et al.*, *J. Phys. Chem. C* **115**, 10851 (2011).
42. M. C. Tan *et al.*, *ACS Appl. Mater. Interfaces* **3**, 3910 (2011).
43. X. Y. Guo *et al.*, *J. Mater. Chem. A* **5**, 20035 (2017).
44. G. P. Li *et al.*, *Inorg. Chem.* **55**, 3951 (2016).
45. M. Zhao *et al.*, *RSC Adv.* **7**, 2258 (2017).
46. P. Sun *et al.*, *J. Colloid Interface Sci.* **547**, 60 (2019).
47. T. T. Wang *et al.*, *Spectrochim. Acta A: Mol. Biomol. Spectrosc.* **254**, 119655 (2021).

Chapter 4

Graphene dot-embedded porous WO$_3$ photoanode for highly efficient photoelectrochemical water splitting

Zhenyan Xu, Xianfeng Zhao, Xiao Xu Yan, Huidan Lu*[‡] and Yongping Liu[†‡]

Guangxi Key Laboratory of Electrochemical and Magneto-chemical Functional Materials
College of Chemistry and Bioengineering, Guilin University of Technology
Guilin 541004, P. R. China
*lhuidangl@163.com
†liuyp624@163.com

The development of semiconductor photoanodes with improved photo-electrochemical (PEC) efficiency and stability for the purpose of realizing solar water splitting is a crucial challenge that carries substantial practical implications.

In this study, GQDs/WO$_3$ composite porous photoanodes were fabricated using a pulsed anodization method in an aqueous electrolyte containing hydroxyl GQDs. The morphology and crystal structure of the as-prepared GQDs/WO$_3$ porous films were characterized by SEM, TEM, XRD, XPS, Raman spectra, and FTIR. The successful embedding of GQDs into the WO$_3$ walls to form tightly bound heterojunctions was confirmed. It was investigated how the GQD content influenced the photoelectrochemical properties of the GQDs/WO$_3$ heterojunctions. The optimal photocurrent densities observed in the

---

[‡]Corresponding authors.

To cite this article, please refer to its earlier version published in the Functional Materials Letters, Volume 16(7), 2340024 (2023), DOI: 10.1142/S1793604723400246.

GQDs/WO$_3$ samples (10-WO$_3$) were found to be three times higher compared to those obtained from pure WO$_3$ porous films. This significant enhancement clearly indicates the positive impact of GQDs on the PEC performance. The boosted PEC activity of the GQDs/WO$_3$ heterojunction photoanode was attributed to the fact that the introduction of GQDs promotes more efficient light absorption, charge separation, and transport processes within the heterojunction structure. Moreover, the GQDs/WO$_3$ composites photoelectrodes exhibited excellent PEC stability, predicting its enormous potential for practical applications.

Keywords: Tungsten oxide; graphene quantum dots; composites photoanode; PEC performance; water splitting.

1. Introduction

Converting solar energy into storable hydrogen fuel via water splitting is one of the most promising approaches for sustainable hydrogen production, which can help meet the ever-growing global energy demand. Due to water oxidation being the rate-limiting step in the process of water splitting, there has been a significant focus on the development of highly efficient photoanodes for the evolution of O$_2$. This area of research has attracted considerable attention in recent times. The challenges in designing efficient heterojunction photoanodes lie in the lagging dynamics of interfacial charge transfer and poor photoelectrochemical stability. Hence, the development of photocatalysts that exhibit low rates of photoinduced charge carrier recombination and exceptional resistance to photo-corrosion is an immensely appealing prospect, albeit one that poses significant challenges.

WO$_3$ possesses excellent corrosion resistance and stability, and its narrow bandgap (Eg = 2.5 eV ~ 2.8 eV) enables it to absorb visible light up to 500 nm, endowing it with enormous potential for application in photoelectrocatalysis.[1] Over the past decade, one-dimensional, two-dimensional, and three-dimensional nanostructured WO$_3$ have been synthesized by various methods. Current techniques for controlling the synthesis of WO$_3$ nanostructures include hydrothermal synthesis,[2] sol-gel method,[3] chemical vapor deposition,[4] radio frequency magnetron sputtering,[5] and anodic oxidation[6] among others. Our research group has made some progress in improving the photocatalytic performance by synthesizing

WO_3 nanostructures with different morphologies. We prepared ultrafine WO_3 nanowires via hydrothermal method, which displayed significantly higher photocatalytic degradation efficiency than bulk WO_3. In addition, we also synthesized WO_3 nanosheet arrays by hydrothermal method and three-dimensional porous WO_3 nanofilms by anodic oxidation, both of which exhibited strong photocatalytic signals.[7,8] Besides, constructing heterojunctions is an effective approach to suppress charge carrier recombination and enhance photocatalytic efficiency. For example, we prepared an oxygen-deficient $H\text{-}BiVO_4/WO_3$ nanosheet array photoanode, which showed further improvement in PEC water oxidation.

Graphene quantum dots (GQDs) represent a novel type of 0D carbon nanomaterial, characterized by dimensions below 10 nm, which exhibit remarkable properties such as size-dependent band gap variation, exceptional electron conductivity, favorable optical properties, and uniform dispersion in polar solvents and water.[9-12] Moreover, GQDs are environmentally friendly and possess strong anti-chemical corrosion and anti-ultraviolet (UV) irradiation capabilities.[11] Given these excellent physical and chemical properties, GQDs have shown significant promise in the areas of solar photocatalysis and photo-electrocatalysis. The incorporation of GQDs as a photosensitizer and electron reservoir in the photocatalytic process has been demonstrated to enhance the photocatalytic and photoelectrocatalytic performance of semiconductors.[13,14] Moreover, the incorporation of graphene quantum dots (GQDs) in composite structures has been demonstrated to extend the lifetime of electron-hole (e^-/h^+) pairs and enhance the visible-light absorption capabilities of TiO_2 nanoparticles. These findings suggest that employing GQDs in composite formations could potentially be a viable strategy for improving the performance of photoelectrochemical (PEC) applications.[15]

Several synthesis techniques have been reported for the preparation of photocatalysts loaded with graphene dots (GQDs), including the mechanical ball milling method,[16] the impregnation method,[17] lyophilization,[18] electrostatic absorption,[19] hydrothermal method[20] and others. The interface interaction of heterojunction photocatalysts is closely associated with the fabrication method employed. For instance, Xuanhua Li's group[21] achieved a strong chemical bridging of GQD/In$_2$S$_3$ hybrids by rationally adjusting the surface S vacancy concentration during the hydrothermal process. The interfacial electron exchange between the GQD and In$_2$S$_3$ was found to increase drastically compared to hybrids with a loose interface. *In situ* growth can reduce the contact resistance between the photocatalytic material and the electrode, and enhance the adhesion between the semiconductor material and the metal substrate, thereby improving the PEC activity and stability of the photoanode.[22] Therefore, it is necessary to explore new manufacturing strategies to obtain graphene quantum dots composite photoanodes in order to improve the interface charge transfer/separation efficiency and heighten the photoelectrical performance.

The nanoarchitecture of a photocatalyst plays a crucial role in enhancing its photocatalytic performance. Jianmei Lu and their research team[12] have recently anchored graphene quantum dots uniformly onto the framework of one-pot-synthesized Bi$_2$WO$_6$/WO$_3$ inverse opals, resulting in a significant improvement in photocatalytic efficiency in the degradation and mineralization of phenol under visible light illumination when compared to pure WO$_3$. The research group led by Chuan Dong[19] utilized electrostatic interactions to immobilize graphene quantum dots (GQDs) onto the surface of TiO$_2$ nanotubes, resulting in enhanced visible-light absorption, improved charge separation and transfer, and excellent photoelectrochemical (PEC) activity. However, there are few reports on the application of GQDs to improve the photoelectrochemical properties of WO$_3$, and the reported photoconversion efficiency is not satisfactory.[23] For instance, the group of Yun Lei[24] prepared WO$_3$/GQDs-H composite materials with a "dot-nanoparticle" structure using a hydrothermal method, but the improvement of photoelectrochemical performance was minor, and the photocurrent was low.

In the present study, we have devised a novel approach, namely pulse anodic oxidation, to fabricate GQDs/WO$_3$ porous hybrids from a W substrate. The incorporation of graphene quantum dots (GQDs) in the hybrid structure allows for their dual functionality as both photosensitizers and electron reservoirs. On one hand, the incorporation of graphene quantum dots (GQDs) into GQDs/WO$_3$ composite films enhances their light absorption in the UV–visible range. This increased light absorption results in a higher number of electron-hole pairs in the mesoporous layer, thus leading to a strengthened photocurrent. On the other hand, GQDs are uniformly embedded in the closely connected WO$_3$ pore walls with full contact, which is conducive to shortening the migration path of photogenerated electrons, preventing carrier recombination, and improving the separation efficiency. Third, pulse anodic oxidation method was utilized to grow in-situ ordered porous GQDs/WO$_3$ composite film, which is tightly bonded to the tungsten substrate and has a large specific surface area, thereby facilitating the generation of larger photoelectrical signals. PEC measurements indicate that the photocurrent of GQDs/WO$_3$ is three times higher than that of unmodified WO$_3$ under simulated sunlight irradiation, and the resistance of the composite photoanode decreases while the carrier density increases significantly. This demonstrates that the composite porous films with a special structure constructed by anodic oxidation method exhibit superior photoelectrochemical performance.

2. Experiment

2.1. *Materials and methods*

2.1.1. *Preparation of GQDs-OH*

Graphene quantum dots (GQDs) were synthesized via the hydrothermal method. Specifically, OH-functionalized GQDs (OH-GQDs) were synthesized through the nitration of 2 g of pyrenes in 160 ml of hot nitric acid at 80°C for 12 h with stirring. The resulting solution was then cooled to room temperature, and approximately 3 g of yellow 1,3,6-trinitropyrene was collected via deionized water suction filter. This was followed by hydrothermal treatment in a 0.2 M NaOH aqueous solution at 200°C for 10 h. After that, the

58 *Z. Xu et al.*

solution was purified via dialysis for two days, obtaining a black OH-GQD solution.

2.1.2. *Preparation of GQDs/WO$_3$ porous nano-films*

All reagents used in this study were of analytical grade and were procured from Shanghai Reagents Company (Shanghai, China). Prior to anodic oxidation, the tungsten foil (85 μm, 99.95%) underwent sonication in acetone, ethanol, and distilled water, and was subsequently dried at 50°C. The resultant clean rectangular tungsten foil was then divided into two parts, each measuring 25 × 10 mm, using opaque tape. One of the segments was used for the anodic oxidation process. Nanoporous WO$_3$ films were synthesized via the process of pulse anodic oxidation. This involved the application of rectangular voltage pulses with a duty ratio of 20% and a frequency of 200 Hz, generated by a function generator, with a pulse voltage of 50 V. The anodic oxidation process employed a tungsten foil as the anode and a stainless steel sheet as the cathode, with an aqueous electrolyte solution consisting of 0.06 M NH$_4$F, 1 M (NH$_4$)$_2$SO$_4$, 100 ml water, and varying volumes (5, 10, 15, 20 ml) of GQDs. The anodization process was carried out for a duration of 30 min, with a periodic anodization voltage being supplied by a micro-arc oxidation power supply (JHMAO-380/20A, Xi'an Jin Tang Material Application Technology Co., Ltd., Xi'an, China). Following this, the as-anodized samples were washed with distilled water and subsequently annealed in air at 450°C for a duration of 3 h, with the temperature being raised at a rate of 5°C min^{-1}. The GQDs/WO$_3$ heterojunction films prepared through the aforementioned process were designated as 5-WO$_3$, 10-WO$_3$, 15-WO$_3$ and 20-WO$_3$, respectively. The porous WO$_3$ film was prepared using the same method but without the presence of GQDs.

2.2. *Sample characterization*

The crystal morphologies and microstructure of the samples were analyzed through Scanning Electron Microscopy equipped with an

Energy-Dispersive X-ray Spectroscopy (EDS) attachment (SEM, JSM-638OLV) and Transmission Electron Microscopy (TEM, JEOL, JEM-2100F) operating at 200 kV. The crystalline phase was identified via powder X-ray diffraction (XRD) using an X'Pert3 Powder diffractometer with Cu Ka radiation ($k = 1.5418$ Å) in the 2θ range of 5–90° at a scanning rate of 0.6565/s. X-ray photoelectron spectroscopy (XPS) measurements were conducted using a monochromatized Al Ka excitation on an ESCALAB 250Xi spectrometer. Fourier Transform infrared spectroscopy (FTIR) was performed using the KBr pellet technique on a Nicolet iS10 FTIR spectrometer (Thermo Fisher). Raman spectra were obtained using a Raman spectrometer (Thermo Fisher Scientific) with a 532 nm laser. UV–vis diffuse reflectance spectra (UV–vis DRS) were recorded on a UV–2450 spectrometer (the Shimadzu Corporation) equipped with an integrating sphere attachment at room temperature within the range of 250–800 nm.

2.3. *Photoelectrochemical measurement*

For the evaluation of the PEC activity, a CHI860D electrochemical system (Shanghai, China) was employed in ambient conditions, with a single-compartment quartz cell containing 0.5 M Na_2SO_4 electrolyte, under the irradiation of a Xe lamp (91192-1000, Newport, USA, 1 Sun AM 1.5 G), or in the dark. GQDs/WO_3 composite porous films on a W substrate was utilized as work electrode, while a saturated calomel electrode (SCE) served as the reference electrode and a Pt plate was used as the counter electrode. The potential measured against SCE was converted to the reversible hydrogen electrode (RHE) scale using the Nernst equation.[25]

$$E(\text{RHE}) = E(\text{SCE}) + 0.059\,{}^{*}\text{pH} + 0.2412\text{V}. \tag{1}$$

3. Results and Discussion

Figure 1 displays the X-ray diffraction (XRD) pattern of the GQDs/WO_3 heterojunction films prepared in this study. The peaks

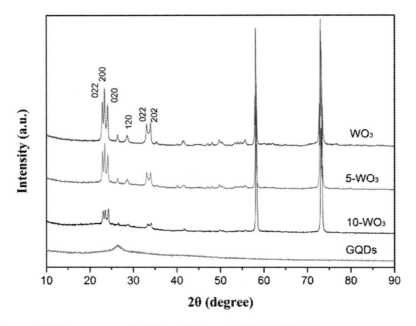

Fig. 1. XRD patterns of GQDs, WO$_3$ and GQDs/WO$_3$ heterojunction nanofilms with different GQD contents.

observed at 22.92°, 23.39°, 24.10°, 28.63°, 33.15° and 33.94° correspond to the (002), (020), (200), (120), (022), and (202) planes, respectively, of monoclinic tungsten oxide (JCPDS data card No. 89-4476). The characteristic diffraction peak of GQDs appears at 26.43°. Compared with pure WO$_3$, as the content of GQDs increases, the diffraction peak position of GQDs/WO$_3$ remains almost unchanged, but the intensity of the diffraction peak gradually weakens. This indicates that the crystal phase of WO$_3$ has not changed after the introduction of GQDs, but an increase in the loading amount of GQDs leads to a decrease in the crystallinity of WO$_3$. Meanwhile, no diffraction peak of GQDs was observed around $2\theta = 26.43°$, which may be due to the low content of GQDs and the weak intensity of its diffraction peak, which was covered by the diffraction peak of WO$_3$.

Figure 2 depicts the scanning electron microscopy (SEM) images of the top and cross-sectional views of WO$_3$ porous films,

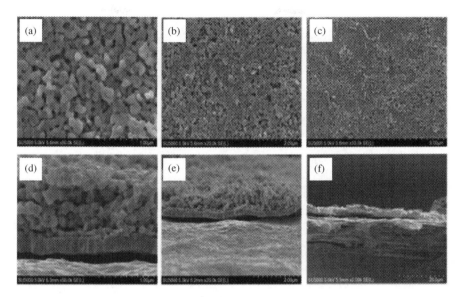

Fig. 2. SEM images of the top section (a, b, c) and cross-section (d, e, f) of WO_3 porous films.

captured at varying magnifications. As shown in Fig. 2(a), many uniform and dense small pores were formed on the surface of the WO_3 film. The cross-sectional image (Figs. 2(d) and 2(e)) shows that the interior of the film also exhibits a porous morphology, with a thickness of about 2 μm. This indicates that the prepared WO_3 film is a three-dimensional porous structure with a large specific surface area, which can promote the separation and migration of interface charges. Figures 2(c) and 2(f) show that large-area porous WO_3 films can be prepared by anodization, which is highly advantageous for the application of WO_3 photoanodes.

Upon closer observation, it was found that the morphology of the GQDs/WO_3 composite porous films is similar to that of the WO_3 films, which implies that the addition of GQDs did not alter the structure of the porous WO_3 films, consistent with the XRD test results. Figure 3 displays the EDS mappings of W, O, and C in the GQDs/WO_3 nanocomposite, indicating that anodization can successfully combine porous WO_3 with GQDs. It is worth mentioning that as shown in Fig. 3(d), the carbon element in the composite

62 Z. Xu et al.

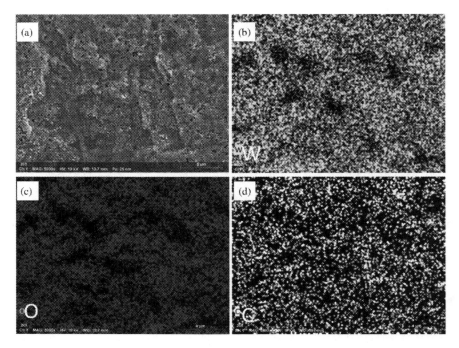

Fig. 3. EDS mappings of GQDs/WO$_3$ composites porous films.

film is evenly distributed, suggesting that GQDs may have penetrated into the interior of the porous WO$_3$ structure and fused with the pore walls of WO$_3$, rather than just being on the surface of WO$_3$. Generally, using freeze-drying,[18] immersion,[15] electrostatic adsorption[19] to modify the surface of semiconductor materials with GQDs can limit the improvement of the migration efficiency of photo-generated carriers because the internal crystals cannot form contact interfaces with GQDs. By using anodization, GQDs can coexist with porous WO$_3$ to form a unique dot-porous network composite structure. The sufficient contact between the two contributes to promoting the separation of photo-generated electrons and holes, shortening the charge transfer path, and enhancing photoelectrochemical efficiency.

To further validate the presence of GQDs in the WO$_3$ composite, transmission electron microscopy (TEM) characterization

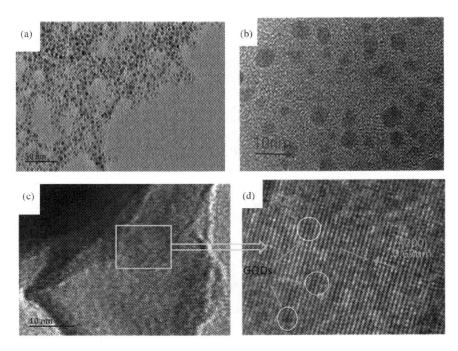

Fig. 4. TEM images of GQDs (a, b) and GQDs/WO$_3$ composites porous films (c, d).

was employed. Figure 4 displays the TEM images of both GQDs and GQDs/WO$_3$ nano-films. From Figs. 4(a) and 4(b), it can be clearly seen that GQDs are distributed in a single-layer scattered manner, and the diameter of a single graphene quantum dot is approximately 3–6 nm (Fig. 4(b)). Figure 4(c) shows the microstructure of the GQDs/WO$_3$ composite film. Upon careful observation, it can be found that there are many small black dots distributed within the WO$_3$ material. The microstructure in Fig. 4(c) was magnified locally to obtain Fig. 4(d). Figure 4(d) clearly shows that the lattice stripe width of the material is 0.67 nm, corresponding to the (200) crystal plane of monoclinic phase WO$_3$.[26] There are many circular dark areas dispersed within WO$_3$, which have the same size and shape as GQDs. The lattice at the bottom can be seen through the dark areas, indicating that the dark areas are ultra-thin graphene quantum dots. This observation suggests that during the growth mechanism of GQDs/WO$_3$ composite films through anodic oxidation of tungsten, the GQDs were effectively integrated with WO$_3$ in a uniform and cohesive manner.

To analyze the elemental composition and surface chemical states of the GQDs/WO$_3$ sample, X-ray photoelectron spectroscopy (XPS) was performed on the sample film. The XPS spectra were calibrated and peak-fitted using the C1s peak at 284.6 eV as a reference. Figure 5(a) shows the XPS survey spectra for the WO$_3$ and GQDs/WO$_3$ samples, which highlight the presence of W, C, and O elements in both samples, with a stronger C peak observed in the GQDs/WO$_3$ sample. Figure 5(b) presents the W 4f spectra for the GQDs/WO$_3$ and WO$_3$ samples, where the peaks located at around 35.58 eV and 37.56 eV can be assigned to the W 4f$_{7/2}$ and W 4f$_{5/2}$

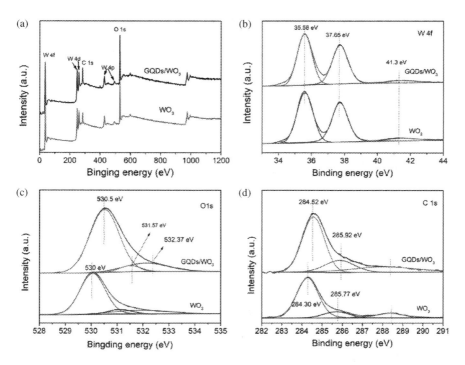

Fig. 5. XPS spectra of WO$_3$ and GQDs/WO$_3$ composites porous films. (a) survey spectrum, (b) W 4f, (c) O 1s, (d) C 1s.

peaks, respectively, which are characteristic peaks of W^{6+}. Figure 5(c) reveals a characteristic peak located at around 530.5 eV in the O 1s XPS spectra of both the GQDs/WO$_3$ and WO$_3$ samples, which is attributed to the chemical binding energy of the W–O bond. In addition, the GQDs/WO$_3$ sample exhibits characteristic peaks located at around 531.57 eV and 532.37 eV in the O 1s XPS spectra, which are attributed to the characteristic peaks of the C=O and O–H bonds in the GQDs. Comparison of the C and O XPS spectra of GQDs/WO$_3$ and WO$_3$ samples reveals a significant shift of the characteristic peak positions towards lower binding energies, which may be attributed to the interaction between the O in WO$_3$ and the C in GQDs, demonstrating the presence of GQDs in the GQDs/WO$_3$ composite. It is speculated that the one-step preparation of the GQDs/WO$_3$ composite film using anodic oxidation method is beneficial for the formation of a tight interface between the two components, which can reduce the interfacial charge transfer resistance.

The presence of GQDs was also confirmed through Raman and infrared spectra. Figure 6(a) displays three prominent peaks at 269, 717 and 807 cm^{-1}, which correspond to the modes of WO$_3$ films, observed in both WO$_3$ and GQDs/WO$_3$ composites. Additionally, two peaks at 1378 and 1592 cm^{-1} were evident in the GQDs/WO$_3$ composite spectra in Fig. 6(a), which can be attributed to the D and G bands of graphite-like materials, respectively. The D band represents the bending vibration peak of sp3 hybrid carbon atom on an aromatic ring, which is indicative of defects and amorphous structure at the edge of the carbon material. The G band, on the other hand, is related to the in-plane vibration of sp^2 carbon atoms, and the intensity ratio of the D band to G band (ID/IG) indicates the degree of graphitization of carbon.[27] These results confirm the presence of GQDs in the GQDs/WO$_3$ composites.

The FT-IR spectra presented in Fig. 6(b) reveal distinct peaks at 3134, 1627, 1396 and 1120 cm^{-1} for the GQDs/WO$_3$ composites. These vibrations are attributed to the OH, C=C, COOH, and C–O–C functional groups, respectively, suggesting the presence of an abundance of oxygenic functional groups in the GQDs. These peak

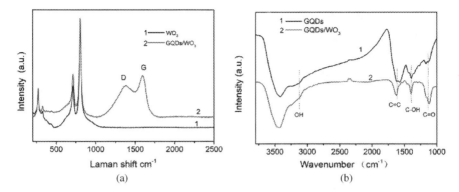

Fig. 6. Raman spectra (a) and FT-IR (b) of WO$_3$ and GQDs/WO$_3$ composites.

locations are consistent with those reported in previous studies on GQDs.

The formation of GQDs/WO$_3$ composites through the C–O bond between GQDs and WO$_3$ was confirmed by analyzing the Raman and FT-IR spectra, in conjunction with the results obtained from XPS measurements.

The UV–vis diffuse reflectance spectra of the porous WO$_3$ film and GQDs/WO$_3$ composite porous films are shown in Fig. 7. The UV–vis absorption edge of the bare WO$_3$ film is located at 450 nm, indicating that it can absorb solar light within 450 nm. However, the absorption edge of the composite films (5-WO$_3$, 10-WO$_3$, 15-WO$_3$ and 20-WO$_3$) loaded with GQDs is red-shifted to 490 nm, verifying that the introduction of GQDs can broaden the visible light absorption range of WO$_3$. Furthermore, the GQDs/WO$_3$ composites exhibit significantly higher absorption values in both the UV and visible regions compared to the pure WO$_3$ film. Notably, the absorption values in the visible region progressively increase with the incorporation of higher GQDs content. Our results are consistent with previous studies, indicating that GQDs modification can significantly enhance the UV–vis absorption ability of semiconductor materials, promote the separation of electron-hole pairs under light illumination, and improve the PEC performance of GQDs composites.[19]

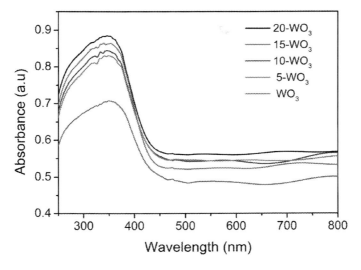

Fig. 7. UV–vis diffuse reflectance spectra of WO$_3$ and GQDs/WO$_3$ heterojunctions film with different contents of GQDs.

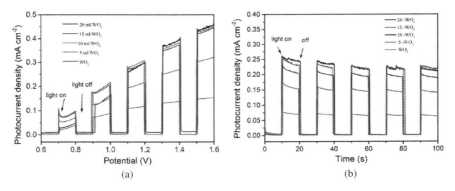

Fig. 8. (a) Liner sweep voltammetry curves and (b) Current-time curves of WO$_3$/GQDs composites porous photoanodes.

In order to evaluate the effect of GQDs on the photoanodic behavior of porous WO$_3$ films, we conducted photoelectrochemical measurements using a three-electrode system in a 0.5 M Na$_2$SO$_4$ electrolyte solution under simulated solar light illumination. Figure 8 illustrates the rapid enhancement of photocurrent responses in the photoelectrode upon illumination, followed by an immediate return to their initial values upon light cessation. This dynamic behavior signifies the exceptional sensitivity of the porous composite film in terms of its photoelectric response.

Simultaneously, a substantial increase in photocurrent is observed upon the introduction of GQDs into the porous WO_3 film. Furthermore, this photocurrent exhibits a progressive enhancement with increasing GQDs doping concentration, as depicted in Fig. 8(a). However, when the amount of GQDs exceeds 10 mL, the photocurrent only slightly increases. It is evident that at an applied potential of 1.6 V, the photocurrent of 10-WO_3 is 0.45 mA/cm^2 versus SCE, which is approximately equal to the photocurrents of 15-WO_3 and 20-WO_3, and three times higher than that of the pristine WO_3 porous film (0.15 mA/cm^2). The significant improvement in the photoelectrochemical properties of GQDs/WO_3 composite material may be attributed to the formation of a closely interfaced heterojunction between WO_3 and GQDs, which promotes charge separation and transfer, inhibits the recombination of photogenerated electron-hole pairs, and thus optimizes the photoelectrochemical conversion efficiency of GQDs/WO_3. The current-time curve of the GQDs/WO_3 composite photoanode was measured at a potential of 0.8 V. As depicted in Fig. 8(b), the trend of change in photocurrent for all samples follows the order of 20-$WO_3 \approx$ 15-$WO_3 >$ 10-$WO_3 >$ 5-WO_3, which is in line with the data obtained from the Linear Sweep Voltammetry (LSV) test. Furthermore, as the loading amount of GQDs increases, the trend of photocurrent attenuation becomes less pronounced, suggesting an improvement in the stability of the electrode.

The capacitance and resistance of electrode materials are indicative of the efficiency of separation and transfer of photo-generated electrons and holes.[28] In order to investigate the oxidation kinetics at the electrode surface, electrochemical impedance spectroscopy (EIS) measurements of the as-prepared films were conducted under visible light illumination with a 0.8 V bias. The typical Nyquist curves of the WO_3 and GQDs/WO_3 heterojunctions films, were obtained over an AC frequency range of 10 kHz–100 MHz, as shown

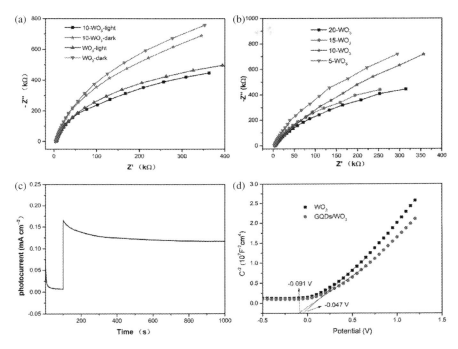

Fig. 9. (a) EIS plots of the bare WO$_3$ and GQDs/WO$_3$ composites photoanodes at a bias of 0.8 V under light and dark, (b) EIS plots of GQDs/WO$_3$ with different volumes of GQDs at a bias of 0.8 V under light and dark, (c) chronoamperometry curve of GQDs/WO$_3$ at a bias of 0.8 V and (d) Mott-Schottky plots of WO$_3$ and GQDs/WO$_3$.

in Figs. 9(a) and 9(b). The Nyquist plot analysis displays that a single capacitive arc is present in all samples, which suggests that Faradaic charge transfer is the rate-limiting step for the oxidation process at the electrode surface. A significant difference in the impedance spectra was observed, as the EIS Nyquist curve of the 10-WO$_3$ film exhibited a smaller circular radius than the bare WO$_3$ film Fig. 9(a). This finding suggests that the 10-WO$_3$ composite film has a lower charge transfer resistance than the WO$_3$ film, implying superior PEC performance. Additionally, Fig. 9(b) shows that the addition of greater amounts of GQDs resulted in smaller arc radii, signifying a faster semiconductor/electrolyte interfacial charge transfer and more effective separation of photogenerated electron-hole pairs in the GQDs/WO$_3$ composites photoanode during light illumination.

Resisting photo-corrosion, reducing photogenerated carrier recombination, and improving photoelectrochemical stability are important challenges for the practical application of photoelectrodes. Figure 9(c) shows the timed current curve of the GQDs/WO$_3$ sample under continuous light illumination at an external voltage of 1.0 V versus RHE. It can be seen that the dark current generated in the first 100 s is very low, and once the light illumination starts, the photocurrent rapidly reaches 0.16 mA/cm^2 and reaches a steady state after 400 s. When subjected to continuous light illumination for a duration of 1000 seconds, the photocurrent value exhibits minimal variation, indicating that the photoelectrode remains in a relatively stable state. The observation suggests that the GQDs/WO$_3$ composite electrode does not experience detachment of GQDs or significant photo-corrosion when subjected to long-term light illumination and immersion in an electrolyte. Therefore, the method of micro-arc anodization can be used to obtain a well-bonded GQDs/WO$_3$ composite film.

In order to investigate the strong correlation between the formation of heterojunctions and the observed enhancement in photocurrent values, Mott Schottky analysis was conducted. The Mott-Schottky plots of WO$_3$ and GQDs/WO$_3$ composite films were obtained under visible light irradiation with a frequency of 1 kHz in a 0.5 M Na$_2$SO$_4$ aqueous electrolyte and presented in Fig. 9(d). Both the bare WO$_3$ and GQDs/WO$_3$ heterojunction films exhibited an n-type characteristic with a positive slope, indicating that electrons served as the majority carriers. The electron density can be calculated based on the Mott-Schottky equation.[29]

$$N_d = \frac{2}{e_0 \varepsilon \varepsilon_0} \left[\frac{\mathrm{d}\frac{1}{C^2}}{\mathrm{d}V} \right]^{-1}.$$

(2)

where ε and ε_0 are the dielectric constant of the semiconductor (50 for WO$_3$[30]) and the vacuum permittivity (8.85×10^{-14} Fcm^{-2}),

e_0 is the electron charge (1.602×10^{-19} C), N_d is the carrier density (cm^{-3}), and V is the potential applied at the electrode, respectively.

The electron density of the WO$_3$ porous film was determined to be 1.77×10^{22} cm^{-3}, which is consistent with previous research findings. However, the carrier density (N_d) of the GQDs/WO$_3$ films was found to be 2.27×10^{23} cm^{-3}, which is one order of magnitude higher than that of the WO$_3$ film. The formation of a heterojunction between GQDs and WO$_3$ thus led to a significant increase in carrier density, which is believed to be a major contributing factor to the pronounced enhancement in photocurrent density observed in the GQDs/WO$_3$ film.

The elevated Nd value observed in the GQDs/WO$_3$ film can be attributed to two factors. Firstly, the GQDs/WO$_3$ porous composite films have a larger interfacial area, which increases the capacitance per unit of electrode area and promotes charge separation and transfer. Secondly, the absorption capability of the GQDs/WO$_3$ composite photoelectrode is strengthened and widened compared to the bare WO$_3$ photoanode, resulting in the production of a higher number of photogenerated electrons. The flat band potential (E_{FB}) at the semiconductor/electrolyte interface can also be estimated using the Motte-Schottky equation.[31]

$$\frac{1}{C^2} = \frac{2}{e_0 \varepsilon \varepsilon_0 N_d} \left[E - E_{FB} - \frac{kT}{e_0} \right]. \tag{3}$$

Where C_{sc} is the capacitance of the space charge layer, E is the electrode applied potential, E_{FB} is the flat band potential, k is the Boltzmann constant, and T is the absolute temperature. The flat band potential (E_{FB}) of the pure WO$_3$ porous film was determined to be -0.091 V versus SCE by extrapolating the x-intercepts of the linear region in the Motte-Schottky plots. In contrast, the E_{FB} of the GQDs/WO$_3$ heterojunction film was found to shift in the negative direction from -0.09 V to -0.047 V. This negative shift of the flat band indicates a decrease in the barrier for oxygen evolution, resulting in a larger accumulation of electrons and better separation of photo-generated carriers in the heterojunction film.

4. Conclusion

In summary, this study for the first time used the anodic oxidation method to prepare GQDs/WO$_3$ composite porous photoanodes on tungsten foil substrates. Structural and morphological characterizations such as XRD, SEM, EDS, and TEM revealed that GQDs entered the porous WO$_3$ structure and were tightly combined with it. XPS, IR, and Raman spectroscopy further confirmed the presence of GQDs in the composite. The introduction of GQDs significantly enhanced the UV–visible light absorption of the GQDs/WO$_3$ heterojunctions and extended its visible light absorption range, reduced the interfacial charge resistance, increased the carrier density, and facilitated the separation and transfer of photogenerated charges in the composites. The results present that the photocurrent of the 10-WO$_3$ porous composite photoanode is three times higher than that of pure WO$_3$ at a potential of 1.6 V versus RHE, suggesting an outstanding improvement in PEC performance. In addition, *in situ* anodization resulted in the uniform distribution of GQDs in the WO$_3$ pore walls, forming a tightly bonded interface, which further enhanced the photoelectrochemical stability of the GQDs/WO$_3$ porous film. Therefore, this study not only demonstrated the important role of GQDs in enhancing the photocatalytic efficiency of semiconductor materials but also proposed a new strategy for preparing heterojunctions photoanodes via anodization.

Acknowledgments

This work was supported by the National Natural Science Foundation of China (22062005, 22165005, U20A20128) and the Guangxi Science Fund for Distinguished Young Scholars (2019GXNSFFA245016). Huidan Lu and Yongping Liu contributed equally to this work.

References

1. G. Hodes *et al.*, *Nature* **261**, 403 (1976).
2. J. S. Jang *et al.*, *J. Korean. Phys. Soc.* **54**, 204 (2009).

3. X. Sun *et al.*, *Appl. Surf. Sci.* **255**, 8629 (2009).
4. C. M. White *et al.*, *Electrochem. Solid. St.* **13**, B120 (2010).
5. R. S. Vemuri *et al.*, *ACS Appl. Mater. Interfaces* **2**, 2623 (2010).
6. F. Fang *et al.*, *Nanotechnology* **22**, 335702 (2011).
7. H. Lu *et al.*, *Microsyst. Nanoeng.* **116**, 17 (2018).
8. H. Lu *et al.*, *J. Solid. State. Electr.* **22**, 2169 (2018).
9. Y. Gao *et al.*, *ChemCatChem.* **10**, 1330 (2018).
10. T. Fei *et al.*, *J. Colloid. Interf. Sci.* **557**, 498 (2019).
11. Z.-d. Lei *et al.*, *J. Hazard. Mater.* **312**, 298 (2016).
12. Q. Zhou *et al.*, *ACS Sustain. Chem. Eng.* **8**, 7921 (2020).
13. H. S. *et al.*, *Int. J. Hydrog. Energy.* **45**, 123 (2020).
14. B. Li *et al.*, *Appl. Surf. Sci.* **478**, 991 (2019).
15. S. Zarei *et al.*, *Eur. Phys. J. Plus.* **136**, 515 (2021).
16. D. K. Chan *et al.*, *J. Environ. Sci.* **60**, 91 (2017).
17. J. Fragoso *et al.*, *Appl. Catal. B-Environ.* **322**, 122115 (2023).
18. Y. Gao *et al.*, *ChemCatChem.* **10**, 1330 (2018).
19. Y. Deng *et al.*, *J. Hazard. Mater.* **426**, 128107 (2023).
20. J. Hou *et al.*, *ACS Appl. Mater. Interfaces* **9**, 24600 (2017).
21. Y. Zhang *et al.*, *Nano Energy* **104**, 107865 (2022).
22. J. Lin *et al.*, *ChemSusChem.* **12**, 961 (2017).
23. D. Jiang *et al.*, *Biosens. Bioelectron.* **183**, 113214 (2021).
24. J. Hu *et al.*, *Mater. Res. Express* **7**, 075602 (2020).
25. W. Liu *et al.*, *ACS Appl. Mater. Interfaces* **14**, 16157 (2021).
26. M. Wang *et al.*, *Energ. Environ. Sci.* **16**, 523 (2023).
27. A. Kelarakis *et al.*, *Curr. Opin. Colloid. In.* **20**, 354 (2015).
28. W. D. Zhang *J. Phys. Chem. C.* **113**, 16247 (2009).
29. F. Zhan *et al.*, *Int. J. Hydrog. Energy.* **40**, 6512 (2015).
30. E. Salje *et al.*, *Acta. Cryst.* **31**, 356 (1975).
31. M. D. Bhatt. *J. Mater. Chem. A* **3**, 10632 (2015).

Chapter 5

High activity and sinter-resistance of Ni@silicalite-1 catalyst for dry reforming of methane

Chunlan Han*, Xiaoxiao Zhu* and Xinping Wang*,[†],[‡]

*State Key Laboratory of Fine Chemical,
Dalian University of Technology,
Dalian 116024, P. R. China
[†]College of Environment and Chemical Engineering,
Dalian University, Dalian 116622, P. R. China
[‡]dllgwxp@dlut.edu.cn

This paper reports a highly active and stable Ni@silicalite-1 catalyst for dry reforming of methane. In the reaction conducted at 800°C, $CH_4/CO_2 = 1/1$ and GHSV = 100,000 mL·g^{-1}·h^{-1} for 100 h, the Ni@silicalite-1 delivered CH_4 conversions of 93% without deactivation. This superior property of the catalyst comes from the special structure that large amounts of Ni (13.7 wt.%) are embedded in the silicalite-1 zeolite crystals as small particles (~4.7 nm), with the zeolite channels transferring molecules while with the zeolite framework resisting the Ni particle migration.

Keywords: Sinter-resistant; highly active; Ni@Silicalite-1; dry reforming of methane.

[‡]Corresponding author.
To cite this article, please refer to its earlier version published in the Functional Materials Letters, Volume 16(7), 2340025 (2023), DOI: 10.1142/S1793604723400258.

Due to thermodynamic restrictions, the dry reforming of methane (DRM) reaction $CH_4 + CO_2 = 2CO + 2H_2$ has to be proceeded at high reaction temperature.[1 4] Ni-based catalysts are believed to be promising candidates due to low cost and superior catalytic activity,[5 7] whereas they suffer from rapid deactivation during the DRM reaction due to metal sintering and carbon deposition via methane cracking ($CH_4 \rightarrow C + 2H_2$) and disproportionation of CO ($2CO \rightarrow CO_2 + C$) over metal surface, in particular over the larger Ni particles.[8] Therefore, how to effectively protect the Ni nanoparticles (Ni NPs) from sintering at the high reaction temperature remains a challenge for catalyst researchers.[9,10]

Encapsulating the Ni NPs into catalyst support has been reported to be a way to resist the Ni NPs sintering.[11 15] Wang et al. reported a 3.6 wt.%Ni@SiO$_2$ catalyst with ~5 nm Ni NPs that displayed a strong sintering-resistance in the DRM reaction at 750°C due to the confinement effect of the silica shell.[11] Bian et al. reported a silica@Ni@silica catalyst with sandwich-like structure, in which multiple small Ni NPs (~6 nm) were supported on the inner silica core but encapsulated within silica shell. The catalyst exhibited much high activity towards the DRM reaction and high stability at 800°C.[15] To boost mass transport within the catalyst structure, some researchers have turned to using zeolites as the support.[16,17] Kong et al. prepared a 0.3 wt.%Ni@S-1 catalyst with the Ni NPs being embedded into Silicalite-1 zeolite (S-1). This catalyst exhibited not only high stability but also much high catalytic activity under the reaction conditions of 650°C, CH_4: $CO_2 =$ 2: 1, GHSV 120,000 mL·h^{-1}·g^{-1} and 5 bars.[18] Recently, Wang et al. synthesized a 0.3 wt.%Ni@S-2 catalyst with ~2.6 nm Ni NPs, and found that the catalyst has superior activity and stability.[19]

Given that the DRM reaction occurs on Ni metal surface,[1,8] higher Ni loading on a support is required to achieve higher mass-specific reaction rate.[20] This paper reports a 13.7 wt.%Ni@silicalite-1 catalyst with Ni NPs size of ~4.7 nm, which has superior activity and stability for the DRM.

The silicalite-1 zeolite (S-1) applied in this work is prepared according to the literature.[21] As represented in Fig. S1, it has a spherical morphology and average size of 100–200 nm.

The Ni@S-1 catalyst was prepared from recrystallization of the S-1 with $Ni(NO_3)_2 \cdot 6H_2O$ in the following procedures: an appropriate amount of $Ni(NO_3)_2$ for a nominal 15 wt.% Ni loading was impregnated on the S-1 by adding one gram of the zeolite to 2 mL $Ni(NO_3)_2$ aqueous solution containing 0.88 g $Ni(NO_3)_2 \cdot 6H_2O$, stirring the mixture at 50°C until drying. Then the resultant materiel was mixed with 40 ml TPAOH (0.15 M) and crystallized at 170°C in autoclave for three days. The obtained solid was collected by centrifugation, and subsequently dried and calcined at 540°C in air for 4 h.

To emphasize the significance of the zeolite channels on mass transfer and the zeolite framework on the Ni NPs resistance to sintering, two comparison catalysts, $Ni@SiO_2$ and Ni/S-1, were prepared. To prepare the $Ni@SiO_2$, a solution obtained by dissolving 4 mmol of $Ni(NO_3)_2 \cdot 6H_2O$ and 11.7 mmol of tetraethyl orthosilicate in 20 ml of ethanol is poured into 20 ml dilute ammonia (3.7 wt.%) and intensively stirred for 10 min and then centrifuged. The precipitate was washed several times with ethanol, dried at 100°C overnight and finally calcined at 540°C in air for 4 h. The Ni/S-1 containing 15 wt.% Ni, was prepared by impregnating the S-1 with $Ni(NO_3)_2$ and calcining the resultant solid at 540°C in air for 4 h.

From ICP-OES analysis, it was known that the Ni content is 13.7 wt.% for Ni@S-1 and 15.5 wt.% for $Ni@SiO_2$.

X-ray powder diffraction (XRD) of the samples was performed on a Rigaku Smartlab nine X-ray diffractometer, which employed Cu Kα radiation ($\lambda = 1.5406$ Å) and operated at 40 kV and 100 mA at scan speed of 8°/min with step of 0.02°. Transmission electron microscopy (TEM) of the samples was performed on JEM F200 field emission transmission electron microscope operated at 200 KV. Scanning electron microscopy (SEM) of the samples was performed on NOVA NanoSEM 450 electron microscopes. FTIR spectra of the samples in KBr pellets coming from 16 scans (1 s per scan) were recorded on a Bruker TENSOR 27 FTIR spectrometer with a resolution of 4 cm^{-1}.

The DRM reaction was carried out in a quartz reactor with an inner diameter of 4.0 mm. 10 mg or 5 mg catalyst (60–80 mesh) being diluted with the same mesh quartz sand (500 mg) was loaded on the reactor and reduced at 750°C with pure H_2 on line for 1 h. After purged by Ar at 800°C, the gas was switched to the reactant gas mixture ($CH_4/CO_2 = 1/1$) to conduct the reaction at GHSV of 100,000 or 900,000 mL·g^{-1}·h^{-1}. The effluent gas was analyzed by an online gas chromatograph (SHIMADZU GC-14A) equipped with a thermal conductivity detector. For the product of DRM reaction over the catalysts, the H_2/CO ratio fell in the range of 0.85–0.90, no large difference was found.

The CH_4 conversion and CO_2 conversion were, respectively, calculated according to the following equations:

$$Conv._{CH4}(\%) = \frac{F_{CH4,\,in} - F_{CH4,\,out}}{F_{CH4,\,in}} \times 100\%, \quad (1)$$

$$Conv._{CO2}(\%) = \frac{F_{CO2,\,in} - F_{CO2,\,out}}{F_{CO2,\,in}} \times 100\%, \quad (2)$$

where $F_{CH4,\,in}$, $F_{CO2,\,in}$, $F_{CH4,\,out}$ and $F_{CO2,\,out}$ denote the flow rates of CH_4 and CO_2 at the inlet or outlet of the reactor.

Figure 1 shows XRD patterns of the Ni@S-1, Ni@SiO$_2$ and Ni/S-1 catalysts prior to reduction. Five diffraction peaks at 7.9, 8.8, 23.1, 23.9 and 24.4° in 2θ characteristic of MFI zeolite

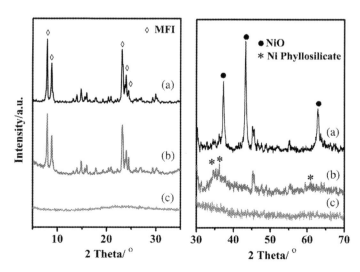

Fig. 1. XRD patterns of the Ni/S-1 (a), Ni@S-1 (b) and Ni@SiO$_2$ (c) catalyst samples prior to reduction.

structure[22] appeared on the Ni/S-1 and Ni@S-1, indicating that the introduction of Ni and recrystallization did not change the crystal structure of S-1. Sharp and strong diffraction peaks at 37.2, 43.1 and 62.8° in 2θ corresponding to the (111), (200) and (220) facets of NiO phase[23] appeared on the Ni/S-1, which indicate low nickel dispersion of sample. No diffraction peak assigned to NiO could be observed on Ni@S-1 and Ni@SiO$_2$. Instead of that, diffraction peaks at 34.1, 36.2, and 60.9° in 2θ, corresponding to the (200), (211) and (060) facets of Ni-phyllosilicate (Ni-P) phase[23,24] appeared on Ni@S-1, while in the case of Ni@SiO$_2$, neither the peaks assigned to NiO nor that assigned to the Ni-P clearly could be observed, only a broad peak in the range of 18–28° ascribed to SiO$_2$ phase could be confirmed.[25]

Figure 2 shows FTIR spectra of the catalysts prior to reduction. Notably, Ni@S-1 exhibited absorption bands at 1036, 710 and 670 cm^{-1} arising from the Ni-P structure,[26-28] while almost no absorption bands appeared on Ni/S-1 as well as on Ni@SiO$_2$, which is very much in line with XRD detection. The results indicate that

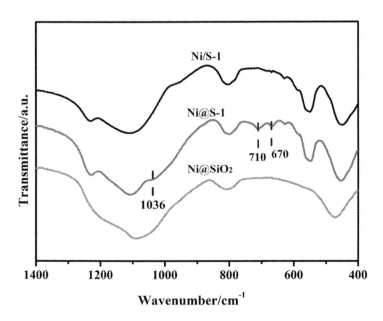

Fig. 2. FTIR spectra of the catalyst samples prior to reduction.

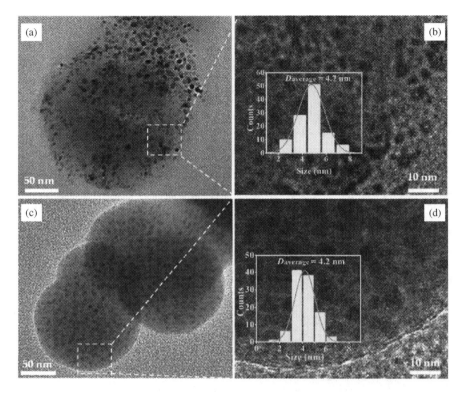

Fig. 3. TEM images of the reduced Ni@S-1 (A, B) and Ni@SiO$_2$ (C, D) catalyst samples.

the preparation process of Ni@S-1 is much more beneficial to the Ni-P structure formation in comparison to that of Ni@SiO$_2$.

Figure 3 shows TEM images of Ni@S-1 and Ni@SiO$_2$ catalysts after reduction by pure hydrogen at 750°C for 1 h. It is clear that Ni is highly dispersed in the substrate for both Ni@S-1 and Ni@SiO$_2$ catalysts. From statistics on the size of 100 particles in TEM images (Figs. 4 and S2), it was known that the average particle size of Ni for the Ni@S-1 and Ni@SiO$_2$ is ~4.7 nm and ~4.2 nm, respectively.

DRM reaction with high space velocity (GHSV = 900, 000 mL· g^{-1}·h^{-1}) was performed on the Ni/S-1, Ni@SiO$_2$ and Ni@S-1 catalysts at 800°C, with which the effect of the catalyst structure on the activity was studied. As shown in Fig. 4, both Ni/S-1 and the

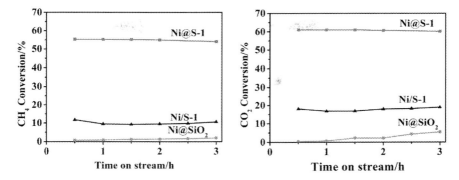

Fig. 4. DRM activities of the catalyst samples (Reaction conditions: GHSV = 900,000 mL·g^{-1}·h^{-1}, CO$_2$: CH$_4$ = 1: 1, T = 800°C).

Ni@SiO$_2$, in particular the latter one, exhibited much lower initial catalytic activity than that of Ni@S-1. For instance, the CH$_4$ conversion is 10% over Ni/S-1, is 1% over Ni@SiO$_2$, while over Ni@S-1, it drastically increased to 55%. For Ni/S-1, the low activity can be ascribed to the much low Ni dispersion, as reflected by XRD (Fig. 1), while for Ni@SiO$_2$, the quite inferior activity is most probably associated with the bad exposure of Ni to the reactant molecules and/or hindered mass transport, as Ni@SiO$_2$ has high Ni dispersion as that of Ni@S-1 (Fig. 3). As for Ni@S-1, the much superior activity is clearly seen from the special structure that Ni has not only high dispersion but also good contact to the reaction molecules being transported by the zeolite tunnels.

To study the durability of Ni@S-1 catalyst, a long-term reaction was carried out over the catalyst at 800°C and GHSV = 100,000 mL·g^{-1}·h^{-1}. As shown in Fig. 5, the initial conversions for CH$_4$ and CO$_2$ were 93%, no deactivation was observed in the whole long-term reaction being conducted for 100 h over the catalyst. The high catalytic stability of Ni@S-1 catalyst can be ascribed to the much strong interaction of nickel with silicon that leads to the N-P formation, as indicated by XRD (Fig. 1) and IR (Fig. 2), as well as that the S-1 framework could act as a barrier obstructing the Ni NPs migration in Coalescence route.[29,31] In the case of Ni/S-1, considerable deactivation was observed within the reaction period of 6 h, resulting from Ni sintering and rapid carbon deposition on

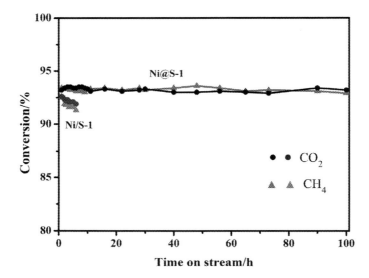

Fig. 5. The conversions of CH$_4$ and CO$_2$ as a function of time on stream for the Ni@S-1 and Ni/S-1 catalyst under the reaction conditions: GHSV = 100,000 mL·g^{-1}·h^{-1}, CH$_4$: CO = 1: 1, T = 800°C.

the large Ni particles because there is no physical confinement hindering Ni NPs migration, as schematically represented by Fig. 6.

For Ni@SiO$_2$, although the Ni is embedded in SiO$_2$ substrate with high dispersion and therefore should possess high catalytic stability, the lower catalytic activity leads to inevitable problems of the catalyst, due to the difficult access of the reactant molecules to the Ni surface resulting from inhibition of the SiO$_2$ shell. As compared in Table S2, Ni@S-1 catalyst has not only much high catalytic activity but also indispensable catalytic stability. Clearly, the superior catalytic property originated from the special structure where Ni is embedded in the S-1 crystals as little Ni particles, and the zeolite framework makes the Ni particles hardly migrate while zeolite tunnels allow the molecules to easily access the Ni active sites. For the much higher sintering-resistance of Ni@S-1 catalyst compared with that of Ni/S-1, apart from the physical obstructing effect of zeolite framework confining the Ni NPs migration, the stronger interaction between nickel and the zeolite support in the case of Ni@S-1 should also have large contribution. As shown in Fig. 7,

High activity and sinter-resistance of Ni@silicalite-1 catalyst   83

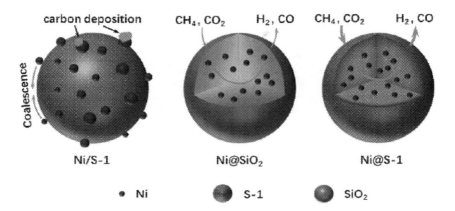

Fig. 6. Schematic illustration for the catalysts with different structures.

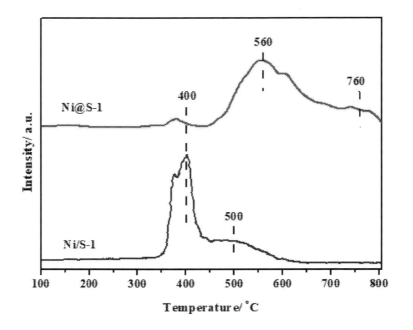

Fig. 7. H$_2$-TPR profiles of the catalyst samples.

almost all of the nickel in Ni/S-1 could be reduced by H$_2$ before 600°C, whereas in the case of Ni@S-1, the reduction has to be prolonged to about 800°C due to stronger interaction between nickel and the zeolite support.

Summarizing this work, Ni@S-1 catalyst possesses obvious advantages on the activity and catalytic durability thanks to the special structure shown in Fig. 6 as well as the stronger interaction between the nickel and the zeolite support.

Acknowledgment

This research was funded by the National Natural Science Foundation of China (grant numbers 22076017 and 21777015).

References

1. N. A. K. Aramouni *et al.*, *Renew. Sustain. Energy Rev.* **82**, 2570 (2018).
2. A. Sternberg, C. M. Jens and A. Bardow, *Green Chem.* **19**, 2244 (2017).
3. Z. Xie *et al.*, *Appl. Catal. B Environ.* **236**, 280 (2018).
4. Y. Wang *et al.*, *ACS Catal.* **8**, 6495 (2018).
5. D. Pakhare and J. Spivey, *Chem. Soc. Rev.* **43**, 7813 (2014).
6. H. Zhou *et al.*, *Appl. Catal. B Environ.* **233**, 143 (2018).
7. Z. Shang *et al.*, *Appl. Catal. B Environ.* **201**, 302 (2017).
8. A. Abdulrasheed *et al.*, *Renew. Sustain. Energy Rev.* **108**, 175 (2019).
9. C. Vogt *et al.*, *ACS Catal.* **10**, 1428 (2020).
10. M. Usman, D. W. M. A. Wan and H. F. Abbas, *Renew. Sustain. Energy Rev.* **45**, 710 (2015).
11. F. Wang *et al.*, *Appl. Catal. B Environ.* **235**, 26 (2018).
12. W. Liu *et al.*, *J. CO_2 Util.* **27**, 297 (2018).
13. Z. Li *et al.*, *ACS Catal.* **4**, 1526 (2014).
14. J. W. Han *et al.*, *ChemSusChem* **7**, 451 (2014).
15. Z. Bian and S. Kawi, *ChemCatChem* **10**, 320 (2018).
16. H. Wang, L. Wang and F. S. Xiao, *ACS Cent. Sci.* **6**, 1685 (2020).
17. S. Kweon *et al.*, *Chem. Eng. J.* **431**, 133364 (2022).
18. W. Kong *et al.*, *Appl. Catal. B Environ.* **285**, 119837 (2021).
19. J. Wang *et al.*, *Appl. Catal. B Environ.* **282**, 119546 (2021).
20. C. Huang *et al.*, *Fuel* **331**, 125957 (2023).
21. L. Zhang, X. Wang and Y. Chen, *Chem. Eng. J.* **382**, 122913 (2020).
22. X. Xie *et al.*, *Int. J. Hydrog. Energy* **48**, 7319 (2023).

23. P. Hongmanorom *et al.*, *Appl. Catal. B Environ.* **282**, 119564 (2021).
24. L. Yan *et al.*, *Green Chem.* **19**, 4600 (2017).
25. Y. Chen, X. Wang and L. Zhang, *Chem. Eng. J.* **394**, 124990 (2020).
26. Y. Lu *et al.*, *Microporous Mesoporous Mater.* **313**, 110842 (2021).
27. X. Kong *et al.*, *ACS Catal.* **5**, 5914 (2015).
28. F. Dong *et al.*, *Catal. Sci. Technol.* **7**, 1880 (2017).
29. T. W. Hansen *et al.*, *Acc. Chem. Res.* **46**, 1720 (2013).
30. K. Wettergren *et al.*, *Nano Lett.* **14**, 5803 (2014).
31. Y. Dai *et al.*, *Chem. Soc. Rev.* **47**, 4314 (2018).

Chapter 6

On the successful growth of bulk gallium oxide crystals by the EFG (Stepanov) method

Dmitrii Andreevich Bauman [*,‡], Dmitrii Iurevich Panov [*], Vladislav Alekseevich Spiridonov [*], Arina Valerievna Kremleva [*] and Alexey Evgenievich Romanov [*,†]

Institute of Advanced Data Transfer Systems, ITMO University
Kronverkskiy pr. 49, St. Petersburg 197101, Russia
†*Regional Engineering Project Office, Togliatti State University*
Belorusskaya St. 14, Togliatti 445020, Russia
‡*dabauman@itmo.ru*

Bulk crystals of β-Ga_2O_3 were successfully grown by the EFG (Stepanov) method. Analysis of the material using an X-ray diffraction showed the high crystalline quality of the obtained crystals. However, when determining the elemental composition by the Energy Dispersive X-ray Spectroscopy (EDS) method, a deviation of the crystal composition from a stoichiometric one was found and a lack of oxygen was detected. Indirectly, this indicates the non-optimal composition of the atmosphere in the growth zone. The origin of this deviation of the composition can be oxygen vacancies in the bulk of the crystal. However, the absorption spectrum does not contain peaks characteristic for oxygen vacancies in gallium oxide. An analysis of the optical transmission spectra made it possible to estimate the optical bandgap of the grown gallium oxide samples, which was 4.7 eV.

Keywords: Gallium oxide; β-Ga_2O_3; EFG; crystal growth; Stepanov method.

---

‡ Corresponding author.

To cite this article, please refer to its earlier version published in the Functional Materials Letters, Volume 16(7), 2340026 (2023), DOI: 10.1142/S179360472340026X.

1. Introduction

In recent decades, wide bandgap semiconductor materials have been of great interest to researchers due to their diverse and important fields of application. An increase in the bandgap makes it possible to use the material for the manufacture of optoelectronic devices for a shorter wavelength region of the spectrum or power electronic devices for high voltages. Two of the first wide bandgap materials to replace silicon were gallium nitride (GaN) and silicon carbide (SiC). The bandgaps E_g of these materials are 3.4 eV for GaN and 3.3 eV for 4H-SiC.[1] However, already at this stage, the fundamental difference between SiC and GaN has manifested the presence of silicon carbide substrates and the absence of such substrates of gallium nitride. In order to manufacture a quality electronic semiconductor device, and in particular a power electronics device where a high breakdown voltage is critical, it is important to ensure the manufacture of low defect epitaxial layers. The most effective way to solve this problem is to use homoepitaxy: growing device structures on "native" substrates. Therefore, a material that makes it possible to produce high-quality and relatively inexpensive substrates will have an advantage.

Today, a potential candidate to replace SiC and GaN is gallium oxide Ga_2O_3, primarily its stable β-form. In terms of the bandgap (E_g = 4.85 eV for β-Ga_2O_3,[2] E_g = 5.3 eV for α-Ga_2O_3),[3] it is second only to AlN, AlGaN, and diamond. Having, in addition, high values of the breakdown field (8 MV/cm theoretically[4] and about 2 MV/cm in the device),[5] high radiation resistance,[6] and relatively high electron mobility (up to 200 $cm^2V^{-1}s^{-1}$),[7] gallium oxide claims to be a new generation material for optoelectronics and power electronics. Gallium oxide also makes it possible to grow bulk crystals by pulling from the melt. This makes it possible to produce substrates for gallium oxide homoepitaxy. Therefore, the choice and development of a method for growing bulk β-Ga_2O_3 crystals play an important role in the development of gallium oxide technologies.

There are at least four widely used methods for growing bulk gallium oxide crystals: the Czochralski method,[8,9] the vertical

Bridgman method,[10] the floating zone method,[11] and the Stepanov[12] or Edge-defined Film-fed Growth (EFG)[13,14] method. Growth features, the design of growth facilities and the physics of the process for each method are described in detail in the book.[15] But it is the EFG method that is actually used today for growing bulk gallium oxide crystals and for manufacturing substrates.

In our work, the EFG method was applied to grow bulk crystals of β-Ga_2O_3. The characteristic dimensions of the grown crystals were about $110 \times 30 \times 4$ mm (L $\times$ W $\times$ T). The elemental composition of the samples was determined, the crystalline quality was analyzed, and the optical absorption spectra were obtained.

2. Materials and Methods

Growth processes were carried out in an industrial growth unit "Nika-3" with induction heating of the crucible. Powdered Ga_2O_3 was used as the starting material. The purity of the material was 99.99% (4N). The crucible and shaper were made from Ir. For seeds, Ga_2O_3 crystallites were used in the form of bars with dimensions of approximately $3 \times 3 \times 30$ mm, cut from previously obtained crystals. The proper choice of a suitable atmosphere in the growth zone plays an important role in growing β-Ga_2O_3 crystals from a melt. The point is that the growth of gallium oxide from a melt is complicated by the chemical instability of this material at the melting temperature[16] and the active evaporation of oxygen from the melt. This requires a special selection of the composition of the atmosphere containing molecular oxygen or in the composition of other compounds (for example, CO_2). Usually, the presence of oxygen in an amount from several units to several tens of volume percent is necessary. In growth experiments, we used the previously selected composition of the atmosphere[17] consisting of a mixture of Ar and O_2. The oxygen content was 4–5% vol, and the pressure in the growth chamber was about 1.1 bar.

The elemental composition of the samples was measured by energy dispersive X-ray spectroscopy (EDS) on a JEOL JSM-7001F scanning electron microscope. This method makes it possible to determine the chemical composition of a sample at a sensitivity level of 0.5 at.%. To analyze the crystalline quality of crystal samples, we used a Dron-8 X-ray diffractometer in a slit configuration with a sharp-focus tube with a copper anode and a NaI (Tl) scintillation detector and a Ni filter for β radiation. The optical properties of the obtained crystals were studied in the wavelength range from 200 nm to 1000 nm using an Avantes Starline AvaSpec 2048 spectrometer. For X-ray and optical measurements, thin (about 0.1 mm thick) plates were fabricated from crystals by cleavage along the (100) cleavage plane. The working plane of the plates was parallel to the crystallographic plane (100). All samples of Ga_2O_3 were entirely beta phase.

3. Results and Discussion

As mentioned above, the dimensions of a typical β-Ga_2O_3 sample grown by the EFG method in our experiments were approximately $110 \times 30 \times 4$ mm (L × W × T). The general view of such a sample is shown in Fig. 1. Visually, the sample is a transparent plane-parallel plate. In Fig. 1(b), a seed crystal is clearly visible on the right end of the sample.

An analysis of the chemical composition of the samples, performed by the EDS method, revealed a deviation from the correct stoichiometric ratio of gallium and oxygen atoms towards a lack of oxygen. A typical picture of the elemental composition of the material is shown in Fig. 2. The specified composition is observed evenly over the entire area of the sample.

As a rule, such a lack of oxygen in the composition indicates the presence of oxygen vacancies in the crystal. Oxygen vacancies presumably affect the optical properties of gallium oxide.[18 20] In particular, according to the results of Ref. 18, the presence of oxygen vacancies increases the absorption in the visible and IR regions of the spectrum. In the same work, it was shown that in the presence of oxygen vacancies, additional peaks appear in the absorption spectrum of gallium oxide in the region from 3 eV to 4 eV. The dependence of the absorption coefficient on the radiation energy for

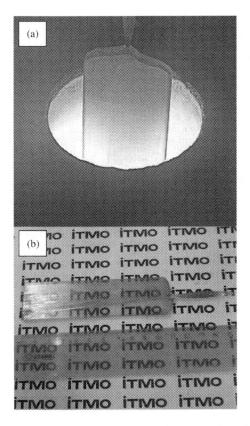

Fig. 1. A sample of a bulk β-Ga$_2$O$_3$ crystal grown by the EFG method: (a) a top view of a growing crystal in the process of growth, (b) a general view of the crystal.

the studied samples of β-Ga$_2$O$_3$ is shown in Fig. 3. It is clearly seen that there are no absorption peaks in the specified range from 3 eV to 4 eV. The same dependence made it possible to estimate the optical bandgap E_g. In a simple approximation for allowed transitions between valence and conductive bands both having parabolic dispersion, the square of absorption coefficient α is proportional to the difference between photon energy $h\nu$ and bandgap energy[21] E_g: $\alpha^2 \sim (h\nu - E_g)$. Thus, bandgap energy can be estimated as an intersection of linear approximation of the dependence α^2 on $h\nu$ with the energy axis (see Fig. 3). The obtained value of E_g is about 4.7 eV.

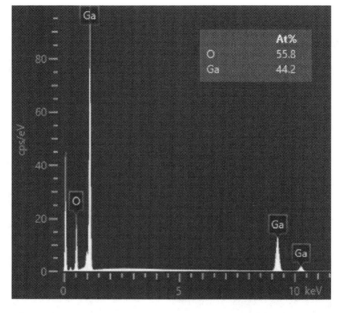

Fig. 2. Bulk crystal composition obtained by the EDS method. The spectra show the mass ratio of elements, the elemental composition in atomic percent is given in the inset in the figure.

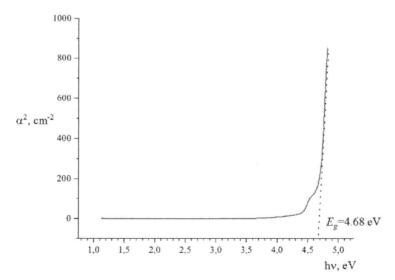

Fig. 3. Optical absorption spectrum: α-absorption coefficient, $h\nu$-radiation photon energy. The dotted line is an approximation of the linear part of the spectrum, which makes it possible to determine the value of the optical bandgap E_g.

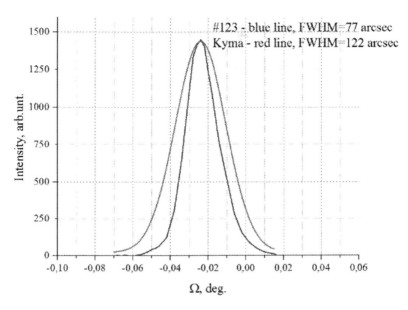

Fig. 4. Results of X-ray diffraction analysis of a bulk β-Ga_2O_3 crystal: blue line — rocking curve Ω for the (800) plane of the sample under study, the FWHM is 77 arcsec; the red line is the Ω rocking curve for the Kyma substrate sample, the FWHM is 122 arcsec.

In the view of the crystal structure, the samples turned out to be of high quality. The rocking curve for reflection from the (800) plane is shown in Fig. 4. The full width at half of height maximum (FWHM) was 77 arcsec. For comparison, the same figure shows a similar rocking curve for a sample of the Kyma substrate, also made by the EFG method. The FWHM for the Kyma substrate is 122 arcsec. The standard 2θ-scan of the Ga_2O_3 sample, shown in Fig. 5, confirms the presence in the crystal of only the β-phase of gallium oxide. Another example for comparison can be samples of bulk Ga_2O_3 crystals grown by the EFG method in the work of Chen et al.[22] The width of the rocking curve for these samples in this work was 67 arcsec. This is comparable with the results obtained in our work.

As can be seen from Fig. 4, the rocking curve is symmetrical and has no shoulders. In the first approximation, this may indicate the absence of twins and/or low-angle grain boundaries in the

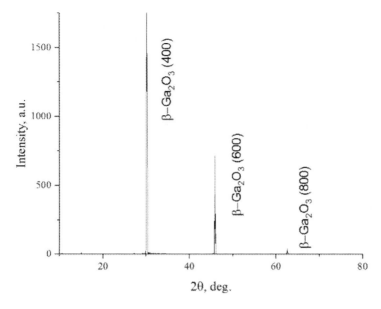

Fig. 5. 2θ-diffraction pattern on the plane (100).

analyzed piece of crystal. The question of the presence of twins, characteristic for bulk gallium oxide crystals, cannot be unambiguously determined only on the basis of X-ray diffraction results.[23] However, this problem can be hardly disclosed within the framework of a short communication and requires a separate studying.

4. Conclusion

Using the Stepanov/EFG method, the samples of bulk crystals of the β-phase of gallium oxide of high crystalline quality were successfully grown. Comparison of the results of X-ray diffraction analysis for the obtained samples with commercially produced substrates allows to state that the obtained material is close to the substrate quality. However, the presence of a composition deviation from the stoichiometric one requires further research. On the one hand, it is necessary to refine the technology and select the oxygen content of the atmosphere, which will provide for this method the suppression of oxygen evaporation, a decrease in the probability of the formation of oxygen vacancies, and the correct stoichiometric composition. On the other hand, it is important to confirm or disprove the presence of oxygen vacancies in the material. For this, it is necessary to understand how oxygen vacancies reveal themselves in the electro-optical properties of a crystal. For example, one can study the absorption spectra of polarized light along different crystallographic directions in the UV region of the spectrum like in Ref. 18. Another way is to analyze the photoluminescence spectra and compare the experimental peaks with those calculated theoretically.[19] This will be the subject of our next work.

Acknowledgments

Authors acknowledge support for this work from the Ministry of Science and Higher Education of the Russian Federation (agreement no. 075-15-2021-1349).

ORCID

Dmitrii Andreevich Bauman ◉ https://orcid.org/0000-0002-5762-5920
Dmitrii Iurevich Panov ◉ https://orcid.org/0000-0001-8715-9505
Vladislav Alekseevich Spiridonov◉https://orcid.org/0000-0001-5751-8597
Arina Valerievna Kremleva ◉ https://orcid.org/0000-0002-7045-0918
Alexey Evgenievich Romanov ◉ https://orcid.org/0000-0003-3738-408X

References

1. J. L. Hudgins *et al.*, *IEEE Trans. Power Electron.* **18**, 907 (2003).
2. S. I. Stepanov *et al.*, *Rev. Adv. Mater. Sci.* **44**, 63 (2016).
3. D. Shinohara and S. Fujita, *Jpn. J. Appl. Phys.* **47**, 7311 (2008).
4. J. Zhang, *et al.*, *APL Mater.* **8**, 020906 (2020).
5. S. Roy *et al.*, *Appl. Phys. Lett.* **122**, 152101 (2023).
6. D. A. Bauman *et al.*, *Acta Astronaut.* **180**, 125 (2021).
7. R. Singh *et al.*, *Mater. Sci. Semicond. Process.* **119**, 105216 (2020).

8. Z. Galazka, *J. Appl. Phys.* **131**, 031103 (2022).
9. D. A. Bauman *et al.*, *Phys. Status Solidi A* **218**, 2100335 (2021).
10. K. Hoshikawa *et al.*, *J. Cryst. Growth* **545**, 125724 (2020).
11. E. Hossain *et al.*, *ECS J. Solid State Sci. Technol.* **8**, Q3144 (2019).
12. P. I. Antonov and V. I. Kurlov, *Prog. Cryst. Growth Ch.* **44**, 63 (2002).
13. A. Kuramata *et al.*, *Jpn. J. Appl. Phys.* **55**, 1202A2 (2016).
14. S. Zhang *et al.*, *J. Semicond.* **39**, 083003 (2018).
15. M. Higashiwaki and S. Fujita (eds.), *Gallium Oxide. Materials Properties, Crystal Growth, and Devices* (Springer, 2020).
16. Z. Galazka *et al.*, *ECS J. Solid State Sci. Technol.* **6**, Q3007 (2017).
17. D. A. Zakgeim *et al.*, *Tech. Phys. Lett.* **46**, 1144 (2020).
18. L. Dong *et al.*, *Sci. Rep.* **7**, 40160 (2017).
19. J. Yao, T. Liu and B. Wang, *Mater. Res. Express* **6**, 075913 (2019).
20. R. Tian *et al.*, *Crystals* **12**, 429 (2022).
21. F. Wooten, *Optical Properties of Solids* (Academic, USA, 1972).
22. B. Chen *et al.*, *CrystEngComm* **25**, 2404 (2023).
23. Y. Bu *et al.*, *CrystEngComm* **25**, 3556 (2023).

Chapter 7

Resistive switching properties of CdTe/CdSe core–shell quantum dots incorporated organic cow milk for memory application

Zolile Wiseman Dlamini ⊕* and Wendy Setlalentoa ⊕[†], Sreedevi Vallabhapurapu ⊕, Tebogo Sfiso Mahule ⊕[‡] and Vijaya Srinivasu Vallabhapurapu ⊕[§], Olamide Abiodun Daramola ⊕[¶] and Potlaki Foster Tseki ⊕[‖], Xavier Siwe-Noundou ⊕ and Rui Werner Macedo Krause ⊕

Department of Maths, Science and Technology Education
Central University of Technology, 20 President Brand St. Bloemfontein
Free State, South Africa
*zdlamini@cut.ac.za
[†]wsetlale@cut.ac.za
School of Computing, University of South Africa
28 Pioneer Ave. Florida Park, Gauteng, South Africa
vallas@unisa.ac.za
Department of Physics, University of South Africa
28 Pioneer Ave. Florida Park, Gauteng, South Africa
[‡]mahults@unisa.ac.za
[§]vallavs@unisa.ac.za
Department of Chemical and Physical Sciences
Walter Sisulu University, Mthatha Eastern Cape, South Africa
[¶]odaramola@wsu.ac.za
[‖]ptseki@wsu.ac.za
Departement of Pharmaceutical Sciences, School of Pharmacy
Sefako Makgatho Health Sciences University
P.O. Box 218, Pretoria 0208, South Africa

---

*Corresponding author.

To cite this article, please refer to its earlier version published in the Functional Materials Letters, Volume 16(7), 2340027 (2023), DOI: 10.1142/S1793604723400271.

98 *Z. W. Dlamini et al.*

xavier.siwenoundou@smu.ac.za
Department of Chemistry, Rhodes University
Grahamstown, Eastern Cape, South Africa
r.krause@ru.ac.za

Our study focuses on the resistive switching memory characteristics of devices containing active layers of CdTe/CdSe core–shell quantum dots (QDs) dispersed in organic cow milk. We fabricated devices containing CdTe/CdSe particles per volume of milk using a direct-dipping method, with particle concentrations of 2.4×10^{-5} (S1), 4.8×10^{-5} (S2), and 7.2×10^{-5} (S3). This method was cost-free. Distinct memory characteristics were observed among devices featuring these concentrations. S1- and S2-based devices exhibited memory behavior with 'S-type' and 'O-type' hysteresis, respectively. The device based on S3 exhibited an initial asymmetric 'N-type' behavior with a large ON/OFF ratio ($\geq 10^{4}$). The memory attribute of the aforementioned device disappeared after the initial three cycles but was subsequently restored by modifying the scan voltage step from 10 mV to 1 mV. The observed results indicate typical symmetric 'N-type' behavior of the device, accompanied by threshold switching under positive voltage bias. Additionally, the switching was observed to be as low as 0.04 V. The S1- and S2-based devices were found to exhibit hopping conduction and Schottky emission in the OFF- and ON-state, respectively, while the S3-based device showed conductive bridge resistive switching as the conduction mechanism. The findings indicate that it is possible to produce biodegradable and disposable memory devices using full cream cow milk and CdTe/CdSe core–shell QDs. The device's switching and memory functions can be manipulated by regulating the quantity of CdTe/CdSe particles present in the milk. Finally, we have demonstrated that the switching behavior of ReRAMs based on milk can be influenced by the voltage steps used during scanning.

Keywords: Cow milk; resistive switching memory; CdTe/CdSe core–shell quantum dots; scan voltage step.

1. Introduction

Studies indicate that despite the increasing usage of electronics, the durability of electronic products has notably decreased.[1] As a result, unused electronic items are frequently discarded, leading to a surge in electric waste (e-waste) that is now a cause for concern.[1,2]

Computer electronics parts, including the chip, motherboard, and memory, often contain hazardous substances like mercury, lead, and cadmium that can be detrimental to living beings, including humans.[3] Efforts are being made to promote the adoption of eco-friendly electronics and green computing to facilitate a shift in the electronics and computing industry.[4] Memory devices are crucial constituents of modern computers. Like other computer components, it is imperative for current memories to decrease their impact on e-waste. Resistive switching memory (ReRAMs) are a developing memory technology that has a simple architecture consisting of an active material, such as a thin film, placed between two electrodes. During the SET/write process or RESET/erase process, ReRAM changes the resistance of the cell from high resistive state (HRS) to low resistive state (LRS) or from LRS to HRS, respectively, in order to store data. The resistive switching (RS) phenomenon was initially observed by Hickmott in 1962.[5] In ReRAM memory, no transistor is required for driving, which allows for limitless miniaturization and a reduced heavy metal footprint compared to current memory giants like dynamic random access memory (DRAM), static random access memory (SRAM), and Flash memory.[6] The discovery of resistive switching in the tobacco mosaic virus[7] has opened up new possibilities for exploring resistive switching (RS) in other biomaterials such as aloe vera,[8] egg albumen,[9] egg-shell,[10] lotus,[11] chitosan,[12 15] silk fibroin,[16] and many more.[17] Despite the progress made in ReRAM research, there remains a dilemma between performance and biocompatibility in ReRAM technology. Thus, further research is required to explore additional RS materials, optimize performance, and enhance our understanding of the switching mechanism in organic-ReRAMs. This will allow organic-ReRAMs to keep pace with the advancements achieved by inorganic-ReRAMs.

This study introduces a ReRAM utilizing full cream organic cow milk containing CdTe/CdSe core–shell quantum dots (QDs). Our current study builds upon our previous findings, which indicated that medium fat bare cow milk displays superior resistive switching compared to full cream milk, which exhibited significant

100 *Z. W. Dlamini et al.*

signal noise.[18] As part of this study, we have incorporated CdTe/CdSe core–shell QDs into full cream milk, resulting in a reduction of the previously observed signal noise and a decrease in the operating voltage of the device. These improvements have led to an enhanced memory window. Furthermore, this scientific article presents a unique discovery of resistive switching in ReRAMs based on cow milk, which is dependent on sweep voltage steps.

2. Experimental Procedure

Daramola *et al.* synthesized, characterized and supplied the CdTe/CdSe core–shell QDs with a particle size of 4.92 nm and a concentration of 2.4 mg.ml^{-1} used in this study.[19] Organic cow milk with full cream was obtained from a local supermarket. We introduced varying volumes, i.e., 0.5 ml, 1.0 ml, and 1.5 ml, of QDs into 50 ml of unprocessed milk. This resulted in concentrations of 2.4×10^{-5} w/v% (S1), 4.8×10^{-5} w/v% (S2), and 7.2×10^{-5} w/v% (S3) of QDs nanoparticles per unit volume (ml) of milk. The QD-milk mixtures were gently stirred for a minimum of 10 min to achieve a uniform distribution of QDs.

The dip-coating method was employed to manufacture the memory devices. Initially, the PET-ITO substrates (Sigma Aldrich, Prod. No. 7497291EA) were subjected to a precleaning process. This involved sonication in acetone, followed by ethanol and ultimately, ultra-distilled water, with each step lasting for 5 min. The substrates were subsequently immersed in the milk+QDs solution for a few seconds. The substrate was delicately extracted and positioned on a level surface with the ITO side facing upwards. It was then allowed to air dry for a duration of 24 h at ambient temperature. Finally, the top electrode (TE) was deposited onto the milk+QDs layers by utilizing a conductive silver (Ag) paste (Sigma Aldrich, Prod. No. 735825). The Ag electrodes were dried at room temperature for more than 48 h. Three devices bearing the notation:

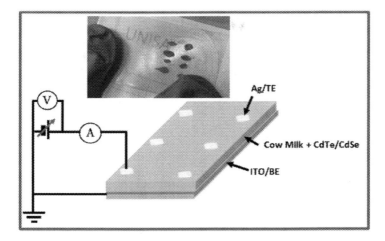

Fig. 1. Schematic diagram of a typical Ag/milk+QDs/ITO device connected to an *I-V* measurement circuit (main figure) and the photograph (inset) of the one of the actual devices.

(i) Ag/milk+QDs(S1)/ITO,
(ii) Ag/milk+QDs(S2)/ITO, and
(iii) Ag/milk+QDs(S3)/ITO,

were fabricated with Ag and ITO as the top and bottom electrodes, respectively. In Fig. 1, the schematic diagram of a typical Ag/milk+QDs/ITO device connected to an *I-V* measurement circuit is presented in the main figure. The insert showcases the photograph of the actual device.

2.1. *Characterization*

To perform characterization, we utilized a scanning electron microscope (SEM) (Zeiss Ultra Plus) and an atomic force microscope (AFM) (Nanosurf FlexAFM) to examine the morphology and topography of the milk films, respectively. We utilized a PerkinElmer LS 55 photoluminescence (PL) spectrometer to measure the fluorescence of plain milk, milk+QDs, and bare QDs with an excitation wavelength of 400 nm. The experiment was carried out at ambient temperature utilizing quartz cells with a path length of 10 nm. The interaction between the QDs and cow milk was further investigated through the use of Fourier transform infrared (FTIR) spectroscopy. The attenuated total reflection method was used to record the IR spectra using the PerkinElmer Spectrum 1 FT-IR spectrophotometer. The frequency range of 4000–650 cm^{-1} was used for each experiment. Finally, the investigation of electricity was carried out utilizing the Keysight Source/Measure Unit (SMU) model B2901A.

3. Results and Discussion

3.1. *Morphology and topography*

The AFM height trace image of organic cow milk mixed with CdTe/CdSe film and deposited on an ITO substrate (as shown in Fig. 2(a)) displays an irregular surface with peaks measuring 160 nm and voids as deep as 148 nm. In this area, the estimated film roughness was 10 nm. The SEM findings (Fig. 2(b)) indicate the presence of fat globules dispersed throughout the film in a random manner. These fat globules, upon desiccation, could have potentially influenced the surface morphology observed in the AFM findings. The AFM phase mapping of the film in 2D (Fig. 2(c)) and 3D (Fig. 2(d)) clearly visualizes sample voids on the topography. These findings are significant as they indicate that the sample is predominantly a single phase. Neither the AFM phase trace nor the SEM images exhibit any evidence of QDs.

3.2. *Photoluminescence spectroscopy*

The interaction between CdTe/CdSe core–shell QDs and cow milk is depicted in the PL spectrum (Fig. 3: main figure). The milk+QDs composite exhibits a blue shift in wavelength, displaying two emission peaks at approximately 523 nm and 570 nm (as shown in Fig. 3, inset (a)). The bare core–shell nanoparticles exhibited a notable decrease in fluorescence intensity. At a wavelength of

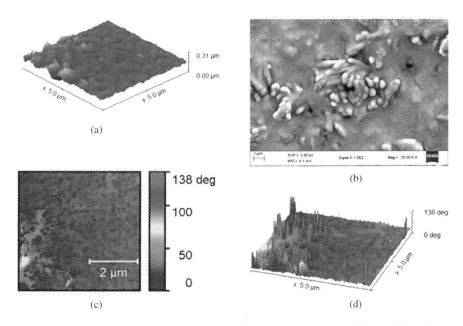

Fig. 2. The 3D AFM height trace image, (a) SEM micrograph, (b) the AFM phase mapping in (c) 2D and (d) 3D for the S3 film deposited on ITO coated PET substrate.

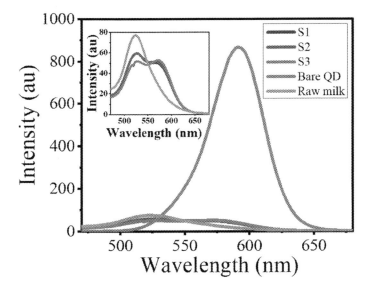

Fig. 3. The PL spectrum of the raw cow milk, raw CdTe/CdSe core–shell QDs and cow milk+CdTe/CdSe core–shell QDs. (a) is the magnified spectra of the raw cow milk and cow milk+DQs, while (b) is the photograph.

590 nm, there was a notable initial increase in fluorescence intensity. The decrease in intensity suggests that energy is being transferred from the fluorescent QDs surface to the non-fluorescent raw milk.[20] The appearance of two emission peaks (at 523 nm and 570 nm) could be attributed to surface reconstruction during the QDs interaction with the raw milk. However, additional research is required to determine the cause of this behavior. The possible mechanism for the interaction between the QDs and milk could be attributed to the electrostatic attraction between the metal ions (Ca^{2+}, Na^+, K^+, Cu^{2+}) existing in milk and the negatively charged carboxylate ions located on the surface of the CdTe/CdSe QDs. In Fig. 3, inset (b), it can be observed that the bare QDs exhibit a bright emission color under UV light, whereas the composite displays lower intensities of emission colors. The reduced emission color intensity of the composite material indicates a decrease in fluorescence intensity, as previously described, resulting from the interaction between the QDs and plain milk. Thus, it can be inferred that these observations suggest a correlation between the CdTe/CdSe QDs and the cow milk.

3.3. *Fourier transform infrared spectroscopy*

The FTIR spectra of the (a) bare QDs, (b) milk+QDs, and (c) plain milk are depicted in Fig. 4. The broad absorption band around 3400 cm^{-1}, which corresponds to the *O–H* stretching, is a characteristic feature of the carboxylic acid group present in the mercaptopropionic and thioglycolic acid capping agents used for the CdTe QDs (a). Additionally, a strong COC stretching absorption band is observed around 1025 cm^{-1}, which is unusual for this frequency range and is likely due to the interaction of the capping agent with the metal. The absorption bands observed at 1600 cm^{-1} and 1594 cm^{-1} are attributed to the bending vibrations of the carbonyl group and alkane bonds, respectively. The spectra obtained from the mixture of QDs and milk showed a reduction in the absorption bands of carboxylic acid groups at 3400 cm^{-1}, which resulted in the

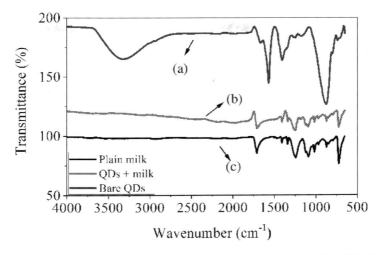

Fig. 4. The FTIR spectra of bare(a) CdTe/CdSe core–shell QDs, (b) the cow milk+CdTe/CdSe core–shell QDs and (c) the plain cow milk.

vanishing of the COC stretch band at 1025 cm^{-1}. Upon interaction with plain milk (b), the carboxyl group's O–H stretching vibration at 3400 cm^{-1} disappeared, as observed. Similarly, the carbonyl group's ($C\!\!=\!\!O$ stretching, 1600 cm^{-1}) absorption band completely disappeared upon interaction of the QDs with plain milk (b).

3.4. *Electric transport study and resistive switching*

The study involved conduction and resistive switching experiments by sweeping the voltage (V) between the Ag and ITO electrodes and simultaneously measuring the current (I) through the active layer. Typically, a device in its original state exhibits insulating properties. As such, we applied high voltage to induce the forming process. The voltage was systematically incremented by 10 mV from zero until either 15 V was reached. This was done to observe any sudden surge in current. The testing of the memory capability of all fabricated devices was conducted by performing a complete cycle voltage scan, which involved going from 0 to $+V_{\text{max}}$, then back to 0, and finally to $-V_{\text{max}}$ and back to 0, while simultaneously measuring the current passing through the device. The compliance current programming of $I_{\rm CC}$ (=100 µA) prevented permanent dielectric breakdown of the devices.

Figures 5(a)–5(c) depict the I-V variation of the Ag/milk+QDs(S1)/ITO device. This device did not show any signs of forming; hence the device was studied at high (15 V) range. In order to provide a lucid depiction of the hysteresis, we have plotted all full cycle scans on a semi-logarithmic scale. Our findings indicate

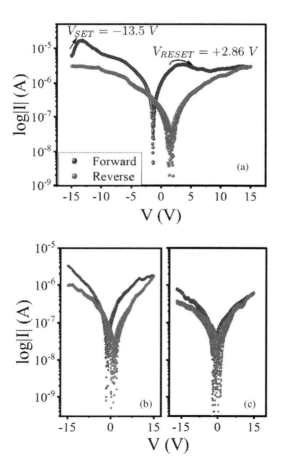

Fig. 5. The semilogarithmic I-V graphs characteristic of the Ag/milk +QDs(S1)/ITO device, during the first (a) second (b) and third (c) voltage scab cycles.

that the flow of current during the forward voltage sweep occurs through a low resistance pathway known as the ON-state, whereas during the reverse sweep, it flows through a high-resistance pathway referred to as the OFF-state. Furthermore, it was observed that the electric current exhibits a non-zero value at 0 V, as previously reported by Guo *et al.*, for the keratin-based organic ReRAM device.[21] They attributed this behavior to ion aggregation at the interface of Ag and keratin film. We believe that similar ion aggregation between Ag electrode and (milk+CdTe/CdSe) film takes place. The transition from a high resistive state (HRS) to a low resistive state (LRS), known as the SET process, appears to take place at −13.5 V. On the other hand, the transition from LRS to HRS, known as RESET, occurs at +2.86 V, as depicted in Fig. 5(a). The observed switching behavior bears resemblance to the symmetric bipolar memory behavior known as the 'S-type', which is commonly observed in scientific literature.[22] The device exhibits an ON/OFF ratio of 22, allowing for clear differentiation between the ON and OFF states during readout. In addition, the voltage scan cycle exhibits a hysteresis range of −15 V to +9.3 V, resulting in a significantly large memory window exceeding 24 V. By sweeping the voltage for the second (Fig. 5(b)) and third (Fig. 5(c)) time, it has been observed that there is a significant increase in signal noise and a decrease in the hysteresis, which in turn affects the ON/OFF ratio.

Figures 6(a)–6(d) show the *I-V* characteristics of the Ag/milk+QDs(S2)/ITO device. Figures 6(a)–6(d) display the first, fifth, and tenth scan cycles of the device, respectively. The hysteresis observed during these cycles exhibits a smooth and homogeneous transition from the ON- to OFF-state and vice versa, indicating an 'O-type' memory behavior. This device has the potential to be used for non-volatile memory applications since there was no significant change in the hysteresis upon sweeping the voltage 10 times. The noteworthy aspect of this device's performance is that the augmentation in QDs concentration led to a decrease in signal noise, resulting in a more consistent and distinct ON/OFF ratio of approximately 10.

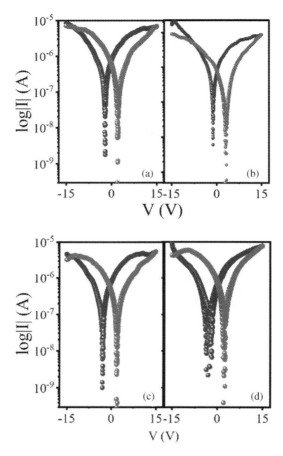

Fig. 6. The semilogarithmic I-V graphs characteristic of the Ag/milk +QDs(S2)/ITO device, during the (a) first, (b) second, (c) fifth, and (d) 10th voltage scan cycles.

Figures 7(a)–7(c) display the I-V characteristics of the Ag/milk +QDs(S3)/ITO device. The data indicates that forming occurs at 4 V, as illustrated in Fig. 7((a) inset). Due to the observed forming process at this voltage, we investigated the memory capacity using a low-voltage amplitude of 2 V. The data presented in Fig. 7(a) of our study demonstrate a notable hysteresis effect exclusively observed under positive voltage bias, suggesting an asymmetric memory behavior. Such behavior has been reported before for other ReRAM systems.[23,24] The I-V hysteresis presents two significant

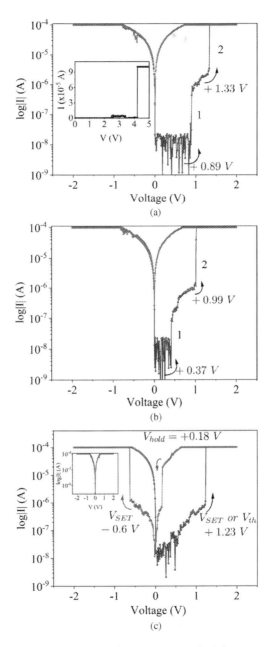

Fig. 7. *I-V* characteristics of the Ag/cow milk+QDs(S3)/ITO device using 10 mV steps. The *I-V* variation during (a) the first cycle, (a: inset) the forming curve, (b) the second cycle, (c) the third cycle and (c: inset) the fourth cycle are depicted.

aspects of interest, namely: (1) the de- vice shows a very large ($\geq 10^4$) ON/OFF ratio, and (2) multilevel switching occurring at low voltages, i.e., $V_{\text{SET}}1 = 0.89$ V and $V_{\text{SET}}2 = 1.33$ V during the SET process. The current curve in the ON-state does not exhibit multilevel switching or any type of switching. Thus, the multilevel switching observed in our device differs from that reported by Sun *et al.* for the EMAR-CNTs composite-based memory with ITO and Al electrodes.[25] Our findings suggest that our device exhibits two distinct HRSs and one LRS. The device's low-switching voltage is crucial for minimizing power consumption in memory devices. Additionally, the ability to switch between multiple levels means that the cell can store multiple bits of data. In the second sweep cycle (Fig. 7(b)), multilevel switching was maintained, but the switching voltages have shifted to lower values (0.37 V and 0.99 V, which is lower than the voltage in the first cycle. In the third cycle, we observed a significant change in the memory behavior, which now exhibits a symmetric "S-type" bipolar behavior as shown in Fig. 7(c). The V_{SET} values are -0.62 V and $+1.23$ V, and the device shows asymmetric threshold switching (TS) at $V_{\text{hold}} = +0.18$ V and $V_{\text{th}} = +1.23$ V, but only under positive voltage bias. The absence of hysteresis in the fourth voltage sweep curve (as shown in Fig. 7(c): inset) suggests that there is no loss of memory behavior.

We modified the voltage sweeping steps from 10 mV to 1 mV, while keeping the step duration constant at 25 ms. Notably, the hysteresis resurfaced, as depicted in Figs. 8(a) and 8(b). The data presented in Fig. 8(a) demonstrate a switching behavior at $V_{\text{SET}} = -0.32$ V and a switching event occurring at $V_{\text{RESET}} = +0.04$ V (see inset) suggesting that the device has transitioned to the 'S-type' bipolar switching mode. Additionally, under the same positive voltage bias, the device continues to exhibit TS behavior with $V_{\text{th}} = +0.32$V and $V_{\text{hold}} = +0.08$V. Upon further scanning of the device, a subsequent alteration in behavior was observed, as depicted in Fig. 8(b). The disappearance of hysteresis in the negative voltage bias has been observed, and a shift in the switching behavior during positive voltage bias has occurred, with switching now occurring at

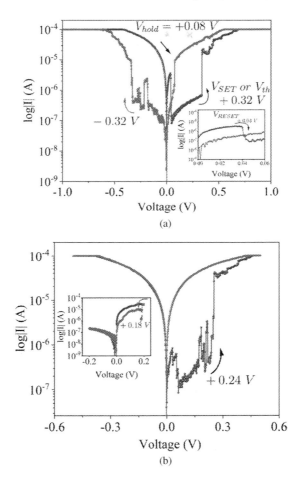

Fig. 8. I-V characteristics of the Ag/cow milk+QDs(S3)/ITO device obtained using 1 mV steps with 25 ms delay. (a) is the first cycle using these new parameters, and the inset shows the magnified 0–0.06 V of the positive region of the graph. (b) displays the second (main image) and third (inset) cycles.

$V_{\text{SET}} = +0.24\text{V}$. The curve for the upcoming scan cycle is depicted in the inset. Based on the graph data, it can be inferred that the device is currently in a state of high resistance (HRS) when subjected to a negative voltage bias. Under positive voltage bias, the current during the forward sweep continues to follow a low-resistance state (LRS) path. However, in the reverse direction transitions from LRS to high-resistance state (HRS) at a voltage of +0.18V. The displayed operating voltage of this device is comparable to the ultra-low voltage (< 0.1 V) observed in the halide perovskite $Ag/CsPbI_3/Pt$, which holds the record for having a large ON/OFF ratio of 10^9.[26] Additional systems, including Cu-doped HfO_2[27] and WSe_2p-FET,[28] among others,[29,30] have exhibited notably low-operating voltages. The effect of altering the scan voltage step is currently inexplicable. As such, it merits a concentrated investigation to gain comprehension.

3.4.1. Conduction mechanism discussion

When voltage stress is applied to the electrodes of a ReRAM cell, a current density may be generated in the active layer, and it follows that $J \propto V^n$ (J is the current density and n is a constant). It is important to consider all possible conduction mechanisms when studying the behavior of dielectric systems like ReRAMs.[31,32] The constant "n" plays a crucial role in determining the dominant conductive mechanism at play. Several conduction mechanisms have been proposed and utilized to explain the behavior of these systems. Therefore, it is essential to carefully analyze each mechanism and its associated parameters to gain a comprehensive understanding of the system's behavior. Each mechanism is characterized by a unique J-V and/or current-temperature (J-T). Table 1 provides a list of the well-known conduction mechanisms used in the literature and their characteristics J-V and J-T relationships.

The I-V curve fitting of the Ag/milk+QDs(S1)/ITO and Ag/milk+QDs(S2)/ITO devices shown in Figs. 9 and 10 depict that the OFF-state current data in both devices can be modeled perfectly with an exponential function, as shown in Figs. 9(a) and 10(b), respectively. The exponential function indicates the hopping conduction mechanism given in Eq. (1):[33]

$$I = I_0 + A \exp\left[\frac{V - V_a}{kT}\right], \tag{1}$$

Resistive switching properties of CdTe/CdSe 113

Table 1. The summary of conduction mechanisms in ReRAM devices.[31]

Name	Voltage dependance	Temperature dependance
Ohmic	$J \propto V$	$\ln J \propto \dfrac{1}{T}$
Space charge limited	$J \propto V$ then $J \propto V^2$	—
Schottky emission	$\ln J \propto V^{\frac{1}{2}}$	$\ln \dfrac{J}{T} \propto \dfrac{1}{T}$
Fowler–Nordheim tunneling	$\ln J \propto V^2$	
Poole–Frenkel emission	$\ln \dfrac{J}{V^2} \propto \dfrac{1}{V}$	$\ln J \propto \dfrac{1}{T}$
Ionic conduction	$J \propto V$	$\ln(JT) \propto \dfrac{1}{T}$
Trap assisted tunneling	$\ln J \propto \dfrac{1}{V}$	—

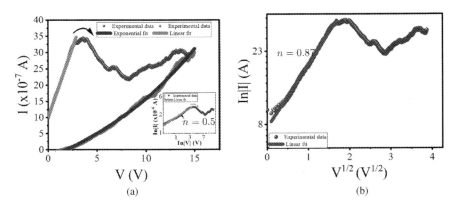

Fig. 9. (a) I-V (main) and $\ln|I|$ vs. $\ln|V|$ (inset) and (b) $\ln|I|$ vs. $V^{(1/2)}$ curve fittings of the Ag/cow milk +QDs(S1)/ITO device.

where I_o, V_a, k, and T are, respectively, the residual current, characteristic voltage, Boltzmann constant, the temperature in Kelvin. At the same time, A is a constant, representing the product of the mean hopping distance, number of electrons and thermal vibrational frequency.[33]

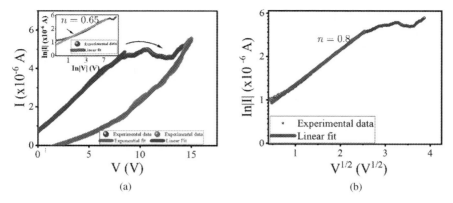

Fig. 10. (a) I-V (main) and ln|I| vs. ln|V| (inset) and (b) ln|I| vs. $V^{(1/2)}$ curve fittings of the Ag/cow milk +QDs(S2)/ITO.

We determined the constant, n, of the ON-state data for both devices by plotting the data on a log–log scale. The values we obtained were 0.5 and 0.65 for the Ag/milk+QDs(S1)/ITO device and Ag/milk+QDs(S2)/ITO device, respectively, as depicted in the insets. Various mechanisms were experimented with and the $\ln I$–$V^{1/2}$ graphs depicted in Figs. 9(b) and 10(b), respectively. Both graphs can be fitted with a linear function with slope of $0.8 \approx 1$. There exists a slight possibility that the conduction mechanism of the ON-state current is Schottky emission. However, further investigation through a temperature-dependent I-V study is required to confirm this possibility. The equation for Schottky emission current is expressed as follows:[31]

$$J = \frac{4\pi m^*(kT)^2}{h^3} \exp\left[-\frac{q\left(\phi_B - \sqrt{\frac{qE}{4\pi\varepsilon}}\right)}{kT}\right]. \quad (2)$$

The symbols m^*, ε, ϕ_B, h, and E in the equation correspond to the effective mass of electrons in the active layer, the permittivity of the active layer, junction barrier height, the Plank's constant, and the electric field across the active layer, respectively. The remaining symbols have their conventional meanings.

We conducted a comparison of the conduction mechanisms of our devices with those of other biological materials-based devices that have been previously reported in scientific literature. The ReRAM device described by Qi *et al.*[11] is based on lactose leaves and utilizes Ag and ITO electrodes. The device contains elements such as K, Ca, Cl, C, and O, which are also present in cow milk.[18] The device exhibits 'O-type' memory hysteresis, and the log–log curve fitting indicates that both the ON- and OFF-state current have $n = 1$, suggesting that the switching is due to the reduction of metallic ions such as Ag^+, Ca^{2+}, or K^+ ions, resulting in conducting filaments. However, our findings do not support the presence of filament conduction. Hence, it is highly probable that the system exhibits a reduced Schottky barrier height at the interface of electrode/milk+QDs. This reduction is further amplified by the rise in applied voltage, resulting in the tunneling of electrons into the active layer. The outcome of our study portrays the Schottky emission current. Reducing the voltage results in a decrease in electron energy, leading to a decrease in the number of electrons that traverse the Schottky barrier. The electrons present in the active layer undergo hopping between the conducting species, including the QDs NP and metal element constituents of the milk. The observed hopping conduction mechanism in the OFF-state of the devices can be attributed to this. It is hypothesized that the utilization of CdTe/CdSe QDs can enhance the quantity of conducting entities present in the active layer, leading to a decrease in current noise.

We have re-plotted the data for Ag/milk+QDs(S3)/ITO on a log–log scale, as depicted in Figs. 11 and 12. Figures 11(a) and 11(b) depict the positive voltage bias replots for Figs. 7(a) and 7(c), respectively. Figures 12(a) and 12(b), on the other hand, display the replots of Fig. 8(a). Based on the displayed graphs, it can be observed that $\ln|I|$–$\ln|V|$ can be represented by a linear function with an approximate slope (n) of 1 during the ON-state. This observation suggests that the flow of current during the ON-state is primarily influenced by charge carriers that are activated by thermal energy. Therefore, the electrical current in this particular state adheres to Ohm's law, which is expressed as follows:

$$I = q\mu N_C \exp\left[-\frac{E_C - E_F}{kT}\right]. \tag{3}$$

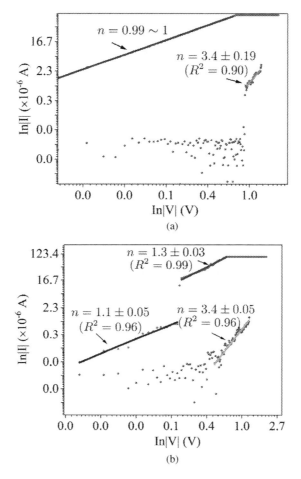

Fig. 11. $\ln|I|$ vs. $\ln|V|$ fitting of Ag/milk +QDs(S3)/ITO, showing the replots of the positive voltage bias of (a) Fig. 6(a) and (b) Fig. 6(c).

The variables q, μ, N_C, E_C, and E_F represent the electronic charge, charge mobility, effective density of state of the conduction band, conduction band, and Fermi level, respectively. The presence of Ohmic conduction during the ON-state indicates that the switching mechanism in this device is a result of the creation of nanoscale conducting filaments that connect the top and bottom electrodes, resulting in a resistive switch. This possibility is supported by the formation behavior depicted in Fig. 7((a) inset). There exist various

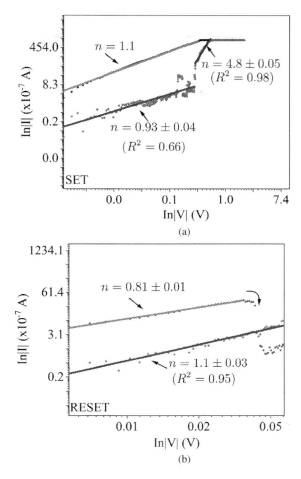

Fig. 12. $\ln|I|$ vs. $\ln|V|$ fitting of Ag/milk+QDs(S3)/ITO, showing the replots of the (a) positive and (b) negative voltage bias of Fig. 8(a).

models for the creation of CFs in conducting bridge ReRAMs. Metallic ions or oxygen vacancies, in the case of metal oxide-based ReRAMs, migrate through the active layer to form CFs. The reduction reaction of these ions is responsible for the formation of CFs. The occurrence of conductive filaments in organic materials can also be attributed to structural damages in the active layer caused by the application of an electric field. The mechanism commonly observed in organic ReRAMs for the formation of CF is the space-charge-limited conduction mechanism (SCLCM). Hence, it is necessary to analyze the behavior of the OFF-state current to gain insight into the potential mechanism of CFs. The data presented in Figs. 11(a) and 11(b) exhibit significant noise in the low-voltage region where the current approaches zero. However, following the initial switch, the behavior depicted in both graphs appears to be similar, with the value of n being approximately 3.4. It is worth noting that there is a relatively large uncertainty range of 0.05–0.19 in both cases. Based on the high value of n, it is possible that this is a case of hopping conduction or a trap-filled state. However, further investigation is needed to confirm either hypothesis. This will be explored in our future research.

Upon modifying the scan voltage steps, our results plotted in Figs. 12(a) and 12(b) continue to demonstrate CF conduction. However, we have observed that the OFF-state current now adheres to a linear variation with $n = 1$. The observed behavior suggests that conductive-bridge switching where the rupture and reformation of conductive filament during the RESET and SET, respectively, takes place. The discovery of altered conduction mechanisms within a single sample is a novel finding that warrants additional research.

4. Conclusion

We investigated the resistive switching behavior of ReRAM devices utilizing CdTe/CdSe core–shell QDs dispersed in cow milk as active layers, which are sandwiched between Ag and ITO. By varying the quantity of QDs present in the milk film, we noted an enhancement in the resistive switching memory performance in terms of both the switching type and behavior. At a high dose of QDs, multiple switching was observed, allowing for the storage of multiple switching which enables the device to store multi-bit of data in a single cell. Also, we observed scan step voltage dependence of both the switching type and switching mechanism. This is a novel observation that requires additional investigation. In summary, our findings indicate that the resistive switching in CdTe/CdSe core–shell QDs dispersed in cow milk is highly dependent on the quantity of nanoparticles present. Additionally, we have determined that the switching and conduction mechanisms in cow milk-based ReRAM can be manipulated by regulating the dispersion of CdTe/CdSe core–shell QDs within the milk's active layer. This is a novel observation that requires additional investigation.

Acknowledgments

Zolile would like to acknowledge the financial support provided by the DHET-nGAP through the Central University of Technology, Bloemfontein and the DST-NRF grant for nGAP Scholars.

Funding Statement

Zolile received funding from the Department of Higher Education and Training-New Generation of Academics Programme (DHET-nGAP) and the Department of Science and Technology-National Research Foundation (DST-NRF) grant for nGAP Scholars. Both these grants were used to support this work.

Conflicts of Interest

The authors declare that they have no conflict of interest to report regarding this study.

ORCID

Z. W. Dlamini ◉ https://orcid.org/0000-0003-0932-0556
W. Setlalentoa ◉ https://orcid.org/0000-0002-2688-8605
S. Vallabhapurapu ◉ https://orcid.org/0000-0002-8737-7980
T. S. Mahule ◉ https://orcid.org/0000-0001-6387-8169
V. S. Vallabhapurapu ◉ https://orcid.org/0000-0001-5341-1442

O. A. Daramola ⊚ https://orcid.org/0000-0002-3793-3811
P. F. Tseki ⊚ https://orcid.org/0000-0002-1056-477X
X. S. Nondou ⊚ https://orcid.org/0000-0002-8667-8351
R. W. M. Krause ⊚ https://orcid.org/0000-0001-6788-6449

References

1. M. K. S. Bhutta, A. Omar and X. Yang, *Econ. Res. Int.* **2011**, 1 (2011).
2. S. Abalansa, B. El Mahrad, J. Icely and A. Newton, *Sustainability* **13**, 1 (2021).
3. W. Nwankwo and P. Uchenna Chinedu, GreenComputing: A machinery for sustainable development in the post-Covid era, in *Green Computing Technologies and Computing Industry, 2021*. Vol. 32 (IntechOpen, 2021), pp. 137–144.
4. P. Pazowski, Green computing: Latest practices and technologies for ICT Sustainability, in *Joint Int. Conf.*, at Bari, Italy, 2015, pp. 1853–1860.
5. T. W. Hickmott, *J. Appl. Phys.* **33**, 2669 (1962). Available at http://aip.scitation.org/doi/10.1063/1.1702530.
6. Y. Li, Q. Qian, X. Zhu, Y. Li, M. Zhang, J. Li, C. Ma, H. Li, J. Lu and Q. Zhang, *InfoMat* **2**, 995 (2020).
7. R. J. Tseng, C. Tsai, L. Ma, J. Ouyang, C. S. Ozkan and Y. Yang, *Nat. Nanotechnol.* **1**, 72 (2006).
8. Z. X. Lim, S. Sreenivasan, Y. H. Wong, F. Zhao and K. Y. Cheong, *Procedia Eng.* **184**, 655 (2017).
9. G. Zhou, Y. Yao, Z. Lu, X. Yang, J. Han, G. Wang, X. Rao, P. Li, Q. Liu and Q. Song, *Nanotechnology* **28**, 425202 (2017).
10. G. Zhou, B. Sun, A. Zhou, B. Wu and H. Huang, *Curr. Appl. Phys.* **17**, 235 (2017).
11. Y. Qi, B. Sun, G. Fu, T. Li, S. Zhu, L. Zheng, S. Mao, X. Kan, M. Lei and Y. Chen, *Chem. Phys.* **516**, 168 (2019).
12. N. Raeis Hosseini and J.-S. Lee, *ACS Nano* **9**, 419 (2015).
13. Z. W. Dlamini, S. Vallabhapurapu, O. A. Daramola, P. F. Tseki, R. W. M. Krause, X. Siwe-Noundou, T. S. Mahule and S. V. Vallabhapurapu, *J. Circuits Syst. Comput.* **31**, 2250113 (16 pages)

(2021) World Scientic Publishing Company DOI: 10.1142/S0218126622501134.

14. Z. W. Dlamini, S. Vallabhapurapu, S. Wu, T. S. Mahule, A. Srivivasan and V. S. Vallabhapu-rapu, *Solid State Commun.* **345**, 114677 (2022).

15. Z. W. Dlamini, S. Vallabhapurapu, O. A. Daramola, P. F. Tseki, R. W. M. Krause, X. Siwe-Noundou, T. S. Mahule and S. V. Vallabhapurapu, *Iran. J. Sci. Technol. Trans. A Sci.* **46**, 709 (2022).

16. X. Ji, L. Song, S. Zhong, Y. Jiang, K. G. Lim, C. Wang and R. Zhao, *J. Phys. Chem. C* **122**, 16909 (2018).

17. K. Y. Cheong, I. A. Tayeb, F. Zhao and J. M. Abdullah, *Nanotechnol. Rev.* **10**, 680 (2021).

18. Z. W. Dlamini, S. Vallabhapurapu and V. S. Vallabhapurapu, *Sustainability* **15**, 8250 (2023), doi:10.3390/su15108250

19. O. A. Daramola, X. Siwe-Noundou, P. F. Tseki and R. W. M. Krause, *Nanomaterials* **11**, 1193 (2021).

20. V. Ghormade, H. Gholap, S. Kale, V. Kulkarni, S. Bhat and K. Paknikar, *J. Biomater. Sci. Polym. Ed.* **26**, 42 (2015).

21. B. Guo, B. Sun, W. Hou, Y. Chen, S. Zhu, S. Mao, L. Zheng, M. Lei, B. Li and G. Fu, *RSC Adv.* **9**, 12436 (2019).

22. D. Prime and S. Paul, *Philos. Trans. R. Soc. A Math. Phys. Eng. Sci.* **367**, 4141 (2009).

23. D. Wu, T. Xu, Z. Shi, Y. Tian, X. Li, Y. Yu and Y. Jiang, *J. Alloys Compd.* **695**, 1653e1657 (2017), doi:10.1016/j.jallcom.2016.10.312.

24. J.-Y. Choi, J. Lee, J. Jeon, J. Im, J. Jang, S.-W. Jin, H. Joung, H.-C. Yu, K.-N. Nam, H.-J. Park, D.-M. Kim, I.-H. Song, J. Yang, S. Cho and C.-M. Chung, *Polym. Chem.* **11**, 7685 (2020), doi:10.1039/D0PY01281E.

25. Y. Sun, J. Lu, C. Ai, D. Wen and X. Bai, *Org. Electron.* **32**, 7 (2016).

26. H. Kim, M.-J. Choi, J. M. Suh, J. S. Han, S. G. Kim, Q. V. Le, S. Y. Kim and H. W. Jang, *NPG Asia Mater.* **12**, 21 (2020).

27. B. Butcher, X. He, M. Huang, Y. Wang, Q. Liu, H. Lv, M. Liu and W. Wang, *Nanotechnology* **21**, (2010).

28. M. Sivan, Y. Li, H. Veluri, Y. Zhao, B. Tang, X. Wang, E. Zamburg, J. F. Leong, J. X. Niu, U. Chand and A. V. Y. Thean, *Nat. Commun.* **10**, 1 (2019).

29. Z. Shen, C. Zhao, Y. Qi, W. Xu, Y. Liu, I. Z. Mitrovic, L. Yang and C. Zhao, *Nanomaterials* **10**, 1 (2020).

30. F. Zahoor, T. Z. Azni Zulkifli and F. A. Khanday, *Nanoscale Res. Lett.* **15**, 90 (2020).
31. E. W. Lim and R. Ismail, *Electron* **4**, 586 (2015).
32. F.-C. Chiu, *Adv. Mater. Sci. Eng.* **2014**, 1 (2014).
33. S. Vallabhapurapu, A. Rohom, N. B. Chaure, C. Tu, S. Du, V. V. Srinivasu and A. Srinivasan, *Appl. Phys. A* **124**, 639 (2018).

Chapter 8

Molecular simulation of (Al–Ga) surface garnished with chromium metal for organic material detecting: A DFT study

Fatemeh Mollaamin ◎* and Majid Monajjemi ◎

*Department of Food Engineering, Faculty of Engineering and Architecture
Kastamonu University, 37100 Kastamonu, Turkey*
Department of Biology, Faculty of Science, Kastamonu University
37100 Kastamonu, Turkey
smollaamin@gmail.com
Department of Chemical Engineering, Central Tehran Branch
Islamic Azad University, 1477893855 Tehran, Iran

Al–Ga surface doped with chromium (Cr) is theoretically studied using first-principles density functional theory (DFT) at the CAM-B3LYP/ EPR-III, LANL2DZ, 6-31+G(d,p) level of theory to explore the chemical adsorption and corrosion inhibition of organic carbenes through coating process. Crystal structure of Cr–(Al–Ga) surface was coated by S–&N–heterocyclic carbenes of benzotriazole (BTA), 2-mercaptobenzo-thiazole (2MBT), 8-hydroxyquinoline (8HQ) and 3-amino-1,2,4-tria-zole-5-thiol (ATR). The NMR spectroscopy of the adsorption of BTA, 2MBT, 8HQ, and ATR on the Cr-doped Al–Ga nanoalloy surface represents that this surface can be employed as the magnetic S–&N–heterocyclic carbene sensors. In fact, Cr site in Cr–(Al–Ga) nanoalloy surface has bigger interaction energy amount from Van der Waals'

---

*Corresponding author.

To cite this article, please refer to its earlier version published in the Functional Materials Letters, Volume 16(7), 2340028 (2023), DOI: 10.1142/ S1793604723400283.

124   *F. Mollaamin & M. Monajjemi*

forces with BTA, 2MBT, 8HQ, and ATR that might cause them large stable towards coating data on the nanosurface. It has been estimated that the criterion for choosing the surface linkage of S and N atoms in BTA, 2MBT, 8HQ, and ATR in adsorption sites can be impacted by the existence of close atoms of aluminum and gallium in the Cr–(Al–Ga) surface. The fluctuation of NQR has estimated the inhibiting role of BTA, 2MBT, 8HQ, and ATR for Cr-doped Al–Ga alloy nanosheet due to S and N atoms in the benzene cycle of heterocyclic carbenes being near the monolayer surface of ternary Cr–(Al–Ga) nanoalloy. Moreover, IR spectroscopy has exhibited that Cr-doped Al–Ga alloy nanosheet with the fluctuation in the frequency of intra-atomic interaction leads us to the most considerable influence in the vicinage elements generated due to inter-atomic interaction. Comparison to ΔG°_{ads} amounts versus dipole moment has illustrated a proper accord among measured parameters based on the rightness of the chosen isotherm for the adsorption steps of the formation of BTA @ Cr–(Al–Ga), 2MBT @ Cr–(Al–Ga), 8HQ @ Cr–(Al–Ga), and ATR @ Cr–(Al–Ga) complexes. Thus, the interval between sulfur, nitrogen, and oxygen atoms in BTA, 2MBT, 8HQ, and ATR during interaction with transition metal of Cr in Cr–(Al–Ga) nanoalloy, (N → Cr, O → Cr, S → Cr), has been estimated with relation coefficient of $R^2 = 0.9509$. Thus, this paper exhibits the influence of Cr doped on the "Al–Ga" surface for adsorption of S–&N–heterocyclic carbenes of BTA, 2MBT, 8HQ, and ATR by using theoretical methods. Furthermore, the partial electron density or PDOS has estimated a certain charge assembly between Cr–(Al–Ga) and S–&N–heterocycles of BTA, 2MBT, 8HQ, and ATR which can remark that the complex dominant of metallic features and an exact degree of covalent traits can describe the augmenting of the sensitivity of Cr–(Al–Ga) surface as a potent sensor for adsorption of BTA, 2MBT, 8HQ, and ATR heterocycles. This work investigates the characteristics, band structure, and projected density of state (PDOS) of Al–Ga nanoalloy doped with Cr element for increasing the corrosion inhibition of the surface through adsorption of organic molecules of carbenes in the surface coatings process. This paper can be helpful in a range of applications which uses Al–Ga alloy for the study of energy storage and adsorption of air pollution or water contamination. Many different approaches such as surface coatings, alloying, and doping can be adopted to protect the surface.

Keywords: Cr–(Al–Ga) surface; BTA; 2MBT; 8HQ; ATR.

1. Introduction

Apart from special needs joint to the real application and the corrosion environment, there are also usual factors to be denoted in material selection.[1] The alloys of metal elements are employed in the branches of car designing, national defense society, airplane industry, electronics and daily significant building with high quality.[2] As a matter of fact, Al material is generated by Hall–Héroult approach with defects of high energy consumption, rigid corrosion of instruments, environmental pollution, and other issues.[3,4]

In fact, heterocyclic organic compounds consisting of multiple bonds and heteroatoms like oxygen, nitrogen, and sulfur are remarkable corrosion inhibitors owing to the absorbance on the metal surface through their heteroatoms. The adsorption of heterocyclic organic compounds on the metal surface closes active zones and decreases the speed of corrosion. Nevertheless, the efficiency of the inhibitor is related to the physico-chemical attributes such as inhibitor structure, being special functional groups, aromaticity, different types of corrosive solutions and charge electron density.[5-8]

It is obvious that heterocyclic carbenes containing S and N atoms are impressible corrosion compounds for diverse series of metals in varied acidic solutions.[9-16] Since various specifications of heterocycles carbenes are remarked, it is crucial that the electrons of a series of heterocycles are considered. Thus, functionalized heterocycles as corrosion inhibiting agents have been selected and identified based on their physico-chemical specifications, and an attention has been assigned to the quantitative assessment of the internal and steric impact of these structures upon their inhibiting output.[17-22] It has been seen that heterocycles containing S and N atoms are appropriate corrosion inhibiting agents for most of metal crystals in diverse acidic conditions. For example, benzotriazole at the Cu and Fe electrode, isoquinoline and imidazole families at the Fe electrode are impressive inhibiting agents for corrosion of these metal crystals.[23-39] The microscopic interaction and reaction mechanism between molecules might be deeply revealed from the quantum chemical specifications which provide an advantageous track to detect adsorption tendency between molecules and interfaces at the atomic and molecular stages.[40-42] Nowadays, it has been shown that pyridine and alkylpyridines had been employed as corrosion inhibiting agents of the Al crystal. There are many investigations that have been devoted to the usage of potent heterocycles and hetero-atomic molecules consisting of organic materials that enhance the anticorrosion qualifications of metal surfaces and alloys. The presence of heteroatoms containing O, S, N, P atoms, aromatic rings and multiple bonds with π-electrons in these inhibiting agents support largely the foundation of inactive blocks on metal crystal and alloys.[43-48] Moreover, it has been analyzed that alkylpyridines have been widely used through their strong polarity which enhances in ionic liquids and solution media.

There is a reasonable connection between the thickness of the passive layer and the effectiveness of inhibiting the corrosion action.[49-51]

In the previous works, the adsorption analysis of pyridine, and nitrogen heterocyclic derivatives onto pristine two-layer aluminum surface,[52] pure monolayer aluminum metal surface based on Freundlich Adsorption[53] has been investigated. Furthermore, the data of Langmuir adsorption model of organic inhibitors containing pyridine and alkylpyridines,[54] BTA, 8HQ, and 2MBT[55] by using monolayer binary alloys of the "Al–Mg", "Al–Ga", "Al–Si" surfaces have been reported. Besides, the role of pyridine and its family compounds as corrosion inhibiting agents for monolayer ternary "Al" nanoalloys including "Al–Mg–Si", "Al–Mg–Ge", Al–Mg–Sn surfaces have been studied.[56]

The interaction of chromium microstructure, and embedding with other atoms applying the density functional theory (DFT) approach, was studied among scientist. The researchers on the basis of Refs. 57 and 58 obtained ultrafine chromium-doped alumina whose luminescence spectrum was identical to that of ruby.[59] Moreover, the spectra of some samples showed additional, unidentified luminescence bands.[60] The spectroscopic properties of Cr-substituted Al_2O_3 single crystals are well known.[61] In the EPR spectra of medium-sized and coarse particles, the resonances due to isolated Cr^{3+} ions differ markedly in strength, whereas the lines of exchange coupled Cr^{3+} ions are close in intensity, which indicates that such particles are close in the amount of chromium incorporated into the lattice of alumina.[60]

Among various approaches to improve the sensing performance of alloy surfaces, the metal-doped method is perceived as effective, and has received great attention and is widely investigated. However, it is still a challenge to construct heterogeneous non-metal/metalloid/metal-doped surface with an excellent sensing performance.

Now, this work intends to extend the previous works[52 56] toward the investigation of Cr-doped Al–Ga surface for adsorption of S–&N–heterocyclic carbenes of benzotriazole (BTA), 2-mercaptobenzothiazole (2MBT), 8-hydroxyquinoline (8HQ), and 3-amino-1,2,4-triazole-5-thiol (ATR) by using through "CAM-B3LYP/EPR-III, LANL2DZ, 6–31+G(d,p)" theoretical methods.

2. Theory, Material and Methods

2.1. *Cr-doped Al–Ga alloy and inhibitory*

The popular kinds of corrosion for Al which are autonomous of the corrosive condition contain pitting, stress-corrosion cracking, exfoliation, intergranular, and galvanic.[62] The non-heat manageable alloys have the larger corrosion persistence against common corrosion compared to the heat manageable alloys. Al-alloys might be sensitive to intergranular corrosion process if second-phase micro components are generated at grain boundary orientations. A corrosion ability of the alloy is separate from that of the matrix will also make intergranular corrosion process. The existence of perceptible values of soluble alloying elements like Cu, Mg, Si, and Zn will cause these alloys sensitive to stress-corrosion cracking process.[63,64] Certain elements like Cu and Mg added to Al alloys ameliorate the mechanical attributes and conduct the alloy to reply the heat dealing. The existence of magnesium also increases resistance and reduces the rate of strength loss at high temperature in the alloys.

Aluminum–Gallium (Al–Ga) which is a degenerate alloy is obtained from liquid gallium diffusing the crystal structure of Ga metal. This alloy is so much fragile that is broken under a little pressure. Al–Ga alloy is also chemically more fragile because it prevents Al from forming a protective oxide layer. This element and its alloys have usually employed experimental fluids for modeling both liquid and solid dynamics in planetary cores. Al–Ga is able to react with H$_2$O molecules to produce Al oxide, Ga metal, and hydrogen gas. Al reacts in air to generate an inactive layer of aluminum oxide and it does not react with water. Aluminum-Gallium alloy can form Al nanoparticles for the hydrogen creating reaction.[65]

In this paper, the adsorption of BTA, 2MBT, 8HQ, ATR as corrosion inhibitors on the Cr–(Al–Ga) surface was coated by S-&N-heterocyclic carbenes of benzotriazole (BTA), 2-mercaptobenzothiazole (2MBT), 3-amino-1,2,4-triazole-5-thiol (ATR), and 8-hydroxyquinoline (8HQ) was assigned by the most suitable Langmuir isotherm, which exhibits the chemisorptive nature of the bond between the inhibitor molecules and the Cr–(Al–Ga) alloy surface (Fig. 1).

2.2. *Multilayer & Multilevel of theoretical ONIOM approach*

From computational level of ONIOM methodology, any combination of three levels in the decreasing order of accuracy might be confirmed in ONIOM3. The especially admissible levels are high

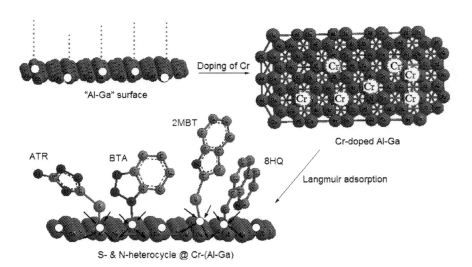

Fig. 1. Introduction of physico-chemical adsorbing properties of S-&N-heterocyclic carbenes of BTA, 2MBT, 8HQ, ATR onto Cr–(Al–Ga) alloy.

QM1, medium QM2, and low QM3 or MM, where high-level QM1 frequently is the *ab-initio* or DFT methodology, medium level QM2 is a low-level *ab initio*, DFT, or semi-empirical (SE) QM methodology, and low level includes an "SE QM" methodology or the "MM" methodology.[66]

Based on this research, ONIOM methodology, QM has been performed due to the DFT methodology of CAM-B3LYP with 6–31+G(d,p) basis set for H, C, S, and O atoms, EPR-III for N atom and LANL2DZ for the transition metals of Cr atom in the adsorption sites. QM2 has been fulfilled on certain Al and Ga atoms in the adsorption sites due to semi-empirical force fields. A QM3 has been discussed on the other Al and Ga atoms with MM2 methodology (Fig. 1).[66]

Therefore, the mixing of three levels of QM, QM2, and QM3 in decreasing the order of validity has been allocated possessing high, medium, and low surfaces of theoretical level.[67]

$$E_{\text{ONIOM}} = E_{\text{high(QM1)}} + E_{\text{medium(QM2)}} + E_{\text{low(QM3)}} \tag{1}$$

ONIOM as a three-layered methodology which authorizes us to unravel a bigger system more explicitly than the one-layered model which may behave like a medium size system stringently like a huge system with usual exactitude (Fig.1).[68] So, the mentioned three-layered sample has been used to activate barriers for the BTA, 2MBT, 8HQ, ATR onto mono-layer Cr–(Al–Ga) alloy surface towards generating the Langmuir adsorption complexes consisting of BTA @ Cr–(Al–Ga), 2MBT @ Cr–(Al–Ga), 8HQ @ Cr–(Al–Ga), and ATR @ Cr–(Al–Ga) (Fig. 1).

2.3. Density Functional Theory (DFT) Method

Our computations have been carried out due to the conceptual DFT using the projector enhanced wave (PAW)[69] methodology. The Perdew–Burke–Ernzerhof (PBE) functional under the generalized gradient approximation (GGA)[70] was applied as the exchange-correlation functional. The nonempirical "PBE" functional is recognized to relegate precise crystal specifications.[71] Commonly, where

GGA functionals lose out, local density functionals stop too. Considering the thermal conductivity compared to local density functionals, PBE does not over attach structures, and therefore the interatomic force constants are not too flexible. All in all, the lattice thermal conductivity is entirely the same whether using local density or GGA functionals.[72 78]

DFT or Density functional theory calculations have been accomplished by using Gaussian 16 revision C.01 program package.[79] The input Z-matrix for the S–&N–heterocyclic carbenes of BTA, 2MBT, 8HQ, ATR as corrosion inhibiting agents adsorbed onto the Cr–(Al–Ga) alloy surface (Fig. 1) has been provided with GaussView 6.1 (Ref. 80). due to the rigid system and coordination format of which a blank line has been cited and using LANL2DZ, EPR-III, 6–31+G(d,p) basis sets to distinguish chemical shielding, frequencies, thermodynamic properties, electrostatic and electronic potential, natural atomic charges, projected density of state and other quantum properties for this work. The rigid PES has been calculated at CAM-B3LYP/EPR-III, LANL2DZ,6–31+G(d,p) for (BTA, 2MBT, 8HQ, ATR) adsorbed onto Cr–(Al–Ga) alloys surface in which the small energy difference between the formations of BTA @ TM–(Al–Ga), 2MBT @ TM–(Al–Ga), 8HQ @ TM–(Al–Ga), and ATR @ TM–(Al–Ga) complexes can direct us to an efficient coated crystal for preventing the corrosion process. Thus, it has been discovered that the crystal binding site preferable of N, O, S-atoms is largely impacted by the essence of neighboring N,O,S-atoms. The calculated (BTA, 2MBT, 8HQ, ATR) @ Ti–(Al–Ga) pair repartition functions have predicted that the formation of crystals addresses to shorter N $\rightarrow$ Cr –alloy, O $\rightarrow$ Cr –alloy, and S $\rightarrow$ Cr –alloy bond lengths when it is figured out to the analogous growth (Fig. 1).

3. Results and Discussion

The adsorption of BTA, 2MBT, 8HQ, ATR as S–&N–heterocyclic carbenes on the Al–Ga Nanoalloy doped with Cr in NaCl solution was allocated by the most appropriate Langmuir isotherm, which remarks the chemisorptive nature of the bond among the (BTA, 2MBT, 8HQ, ATR) @TM–(Al–Ga) complexes, the equilibrium

Molecular simulation of (Al–Ga) surface 131

distribution of ions of the adsorbing compound between the solid and liquid phases, and a monolayer attribute. The adsorbed molecules are kept on the Cr–(Al–Ga) surface with chemisorbed inhibitors having high protection (Fig. 1).

3.1. The properties of electronic specification

The electronic structures of BTA, 2MBT, 8HQ, ATR adsorbed on the Cr-doped Al–Ga surface have been investigated to simplify subsequent study for interfacial electronic properties using CAM-B3LYP/LANL2DZ, 6–311+G (d,p) basis sets.

Figures 2(a)–2(d) show the projected density of state (PDOS) of the (BTA, 2MBT, 8HQ, ATR) @ Cr–Al–Ga surface. The appearance

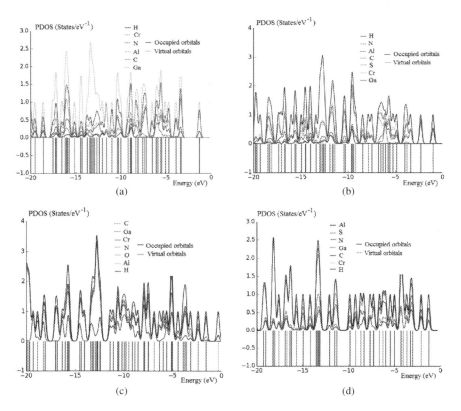

Fig. 2. PDOS adsorption of (a) BTA @ Cr–(Al–Ga), (b) 2MBT @ Cr–(Al–Ga), (c) 8HQ @ Cr–(Al–Ga), and (d) ATR @ Cr–(Al–Ga).

of the energy states (d-orbital) of Cr within the gap of Al–Ga surface induces the reactivity of the system. It is clear from the figure that after doping with Cr atom, there is a significant contribution of Cr d-orbital in the unoccupied level. Based on the population analysis and DOS, it can be concluded that Cr remains in the cationic state and it can accept more electrons from other atoms. Therefore, the curve of partial DOS (PDOS) has described that the p states of the adsorbing process of S & N atom of on the BTA, 2MBT, 8HQ, ATR adsorbed on the Cr–(Al–Ga) surface are overcoming due to the conduction band [Figs. 2(a)–2(d)]. A distinguished metallic trait might be seen in BTA, 2MBT, 8HQ, ATR @ Cr–(Al–Ga)" crystal because of the potent interaction between the p states of C, N, O, S, Al, Ga and the d state of Cr near the Fermi energy. Furthermore, the essence of covalent traits for these clusters has displayed the similar energy value and image of the PDOS for the p orbitals of C, N, O, S, Al, Ga and d orbitals of Cr [Figs. 2(a)–2(d)].

Figures 2(a)–2(d) show that the BTA @ Cr–(Al–Ga), 2MBT @ Cr–(Al–Ga), 8HQ @ Cr–(Al–Ga), and ATR @ Cr–(Al–Ga) states, respectively, have the most contribution at the middle of the conduction band between –5 and –20 eV. Contribution of gallium states in BTA @ Cr–(Al–Ga) complex is enlarged, but hydrogen, carbon, nitrogen, aluminum, and chromium states have little contributions and are similar together [Fig. 2(a)]. Furthermore, contribution of hydrogen, carbon, nitrogen, sulfur, aluminum, and chromium in 2MBT @ Cr–(Al–Ga) complex is enlarged and similar together [Fig. 2(b)]. Moreover, contribution of hydrogen, carbon, oxygen, aluminum, and chromium in 8HQ @ Cr–(Al–Ga) complex is enlarged and similar together, but nitrogen states have little contributions [Fig. 2(c)]. Besides, contribution of gallium states in ATR @ Cr–(Al–Ga) complex are enlarged, but hydrogen, carbon, nitrogen, sulfur, aluminum, and chromium states have little contributions and similar together [Fig. 2(d)].

The partial electron density or PDOS can also estimate a certain charge assembly between Cr–(Al–Ga) and S–&N–heterocycles of BTA, 2MBT, 8HQ, and ATR. On the other hand, the above data can illustrate that the complex dominant of metallic features and an exact degree of covalent traits can describe the augmenting of the sensitivity of Cr–(Al–Ga) surface as a potent sensor for adsorption of BTA, 2MBT, 8HQ, and ATR heterocycles [Figs. 2(a)–2(d)].

3.2. *Magnetism and atomic charge analysis*

Nuclear magnetic resonance spectrum of Cr-doped Al–Ga alloy surface as the potent sensor for adsorbing the S–&N–heterocycles of BTA, 2MBT, 8HQ, and ATR can unravel the role of the indicated transition metal of Cr in the active site of Al–Ga alloy surface through the formation of the covalent binding between S–&N–heterocyclic carbenes (adsorbate) and surface (adsorbent).

From the DFT calculations, it has been attained the chemical shielding $\langle\langle CS\rangle\rangle$ tensors in the principal axes system to estimate the isotropic chemical-shielding $\langle\langle CSI\rangle\rangle$ and anisotropic chemical-shielding $\langle\langle CSA\rangle\rangle$.[81]

$$\sigma_{iso} = \frac{\sigma_{11} + \sigma_{22} + \sigma_{33}}{3}, \tag{2}$$

$$\sigma_{aniso} = \sigma_{33} - (\sigma_{22} + \sigma_{11})/2. \tag{3}$$

The chemical shielding extracts from Nuclear magnetic resonance or NMR can be applied for allocating the structural and geometrical specifications of materials. As a matter of fact, Gauge Invariant Atomic Orbital or GIAO methodology has been recommended as a valid methodology for NMR parameter computations and ONIOM has caught much regards for gaining NMR chemical shielding of inhibitor-surface complexes such as isotropic chemical shielding appears in Eq. (4):

$$\sigma_{iso,ONIOM} = \sigma_{iso,high(QM1)} + \sigma_{iso,medium(QM2)} + \sigma_{iso,low(QM3)}. \tag{4}$$

The NMR data of isotropic σ_{iso} and anisotropic shielding tensor σ_{aniso} for BTA @ Cr–(Al–Ga), 2MBT @ Cr–(Al–Ga), 8HQ @ Cr–(Al–Ga), and ATR @ Cr–(Al–Ga) complexes have been computed by "Gaussian 16 revision C.01" program package[79] and been drawn in Fig. 3.

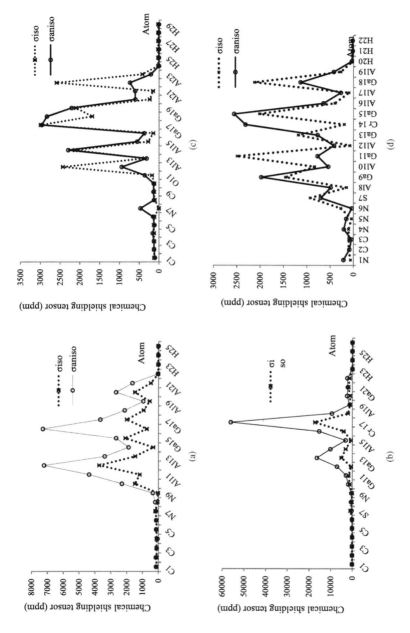

Fig. 3. The NMR spectroscopy results of isotropic shielding tensor (σ_{iso}) and anisotropic shielding tensor (σ_{aniso}) of different elements in S- & N- heterocycles adsorbed on the Cr-doped Al–Ga: (a) BTA @ Cr–(Al–Ga), (b) 2MBT @ Cr–(Al–Ga), (c) 8HQ @ Cr–(Al–Ga), and (d) ATR @ Cr–(Al–Ga) complexes accompanying CAM-B3LYP/EPR-III, LANL2DZ, 6-31+G(d,p).

Molecular simulation of (Al–Ga) surface   135

In fact, the adsorption of BTA, 2MBT, 8HQ, and ATR heterocycles introduces spin polarization on the Cr-doped Al–Ga nanoalloy surface which indicates that this surface might be applied as magnetic sulfur and nitrogen heterocycle detectors. In fact, it is revealed that the isotropic and anisotropy shielding augment with the occupancy in BTA, 2MBT, 8HQ, and ATR heterocycles penetrated by S– and N–atoms in the benzene ring diffusing onto Cr-doped Al–Ga surface [Figs. 3(a)–3(d)].

In Figs. 3(a)–3(d), chromium in the complexes of BTA @ Cr–(Al–Ga) [Fig. 3(a)], 2MBT @ Cr–(Al–Ga) [Fig. 3(b)], 8HQ @ Cr–(Al–Ga) [Fig. 3(c)], and ATR @ Cr–(Al–Ga) [Fig. 3(d)] denotes the fluctuation in the chemical shielding tensor.

In fact, Figs. 3(a)–3(d) indicate the maximum fluctuation for gallium atoms in the neighboring of chromium in the center of metal alloy (adsorbent) for receiving the electrons of nitrogen (N $\rightarrow$ Cr), oxygen (O $\rightarrow$ Cr), and sulfur (S $\rightarrow$ Cr) of BTA, 2MBT, 8HQ, and ATR heterocycles (adsorbate).

3.3. *Electrostatic properties and NQR*

In this research, the calculated nuclear quadrupole resonance or NQR specifications extract from electrostatic properties have been calculated for BTA, 2MBT, 8HQ, and ATR which is in accord to the results of the nuclear quadrupole moment, a trait of the nucleus, and the electric field gradient or EFG in the neighborhood of the nucleus.

As the EFG at the citation of the nucleus in N-heterocycles is allocated by the valence electrons twisted in the particular attachment with close nuclei of Cr-doped Al–Ga alloy crystal, the NQR frequency at which transitions occur is particular for an (N-heterocycles)@(Cr–Al–Ga) complexes (Fig. 4). In NMR, nuclei with spin $\geq 1/2$ have a magnetic dipole moment so that their energies are split by a magnetic field, permitting resonance sorption of energy dependent on the Larmor frequency; $\omega_L = \gamma B$, where γ is the gyromagnetic ratio and B is the magnetic field external to the nucleus. In NQR, nuclei with spin ≥ 1, there is an electric quadrupole

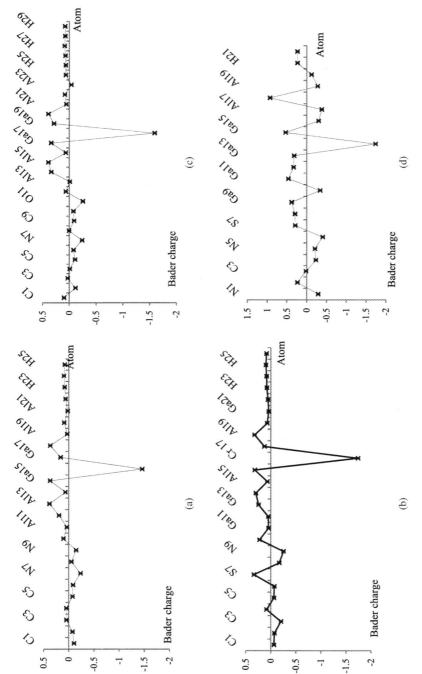

Fig. 4. Bader charge (a.u.) of H, C, N, O, Al, Ga, and Cr atoms in the adsorption process of BTA, 2MBT, 8HQ, and ATR on the Cr-doped Al–Ga alloy by CAM-B3LYP/EPR-III, LANL2DZ, 6-31+G(d,p).

moment which is accompanied with non-spherical nuclear charge diffusions. So, the nuclear charge diffusion extracts from that of a sphere as the oblate or prolate form of the nucleus.[82,83] NQR is a straight frame of the interaction of the quadrupole moment with the EFG which is produced by the electronic structure of its ambiance. Therefore, the NQR transition frequencies are symmetric to the electric quadrupole moment of the nucleus and a measurement of the strength of the local EFG: $\omega \sim \frac{e^2 Qq}{\hbar} = C_q$, where q is dependent on the biggest fundamental portion of the "EFG" tensor at the nucleus, and C_q is the quadrupole coupling constant parameter.[84,85]

In this research work, the electric potential as the quantity of work energy through carrying over the electric charge from one position to another position in the essence of electric field has been shown for BTA@ Cr–(Al–Ga), 2MBT@Cr–(Al–Ga), 8HQ@Cr–(Al–Ga), and ATR@ Cr–(Al–Ga) complexes using "CAM-B3LYP/EPR-III,LANL2DZ, 6–31+G(d,p)" level of theory (Fig. 4).

Furthermore, in Figs. 4(a)–4(d), it has been sketched the Bader charge for some atoms of hydrogen, carbon, nitrogen, oxygen, aluminum, gallium, and chromium in the adsorption process of BTA, 2MBT, 8HQ, and ATR on the Cr-doped Al–Ga alloy surface which have been calculated by CAM-B3LYP/EPR-III, 6–311+G (d,p), LANL2DZ.

In Figs. 4(a) and 4(d), it has been described the influence of the replacement of aluminum metal elements in Al–Ga surface with Cr. It is vivid that the curve of Al–Ga is waved by the transition metal. The sharpest peaks for Bader charge have been shown around transition metal doping of the Al–Ga which present the electron accepting characteristics of chromium versus nitrogen and sulfur atoms of heterocycles in BTA, 2MBT, 8HQ, and ATR [Figs. 4(a)–4(d)].

It has been exhibited that the Bader atomic charge graph of S– and N–atoms has the same tendency; however, a considerable deviation from transition metal element of Cr [Figs. 4(a)–4(d)].

In fact, Figs. 4(a)–4(d) indicate that the gap Bader charge between N, S, O atoms and transitions metal Cr has the maximum value. On the other hand, it can be considered that the efficiency of electron accepting for the Cr-doped on the Al–Ga surface

138 *F. Mollaamin & M. Monajjemi*

indicates the power of covalent bond between nitrogen, oxygen, sulfur, and chromium.

The amounts of changes of charge density have exhibited a more important charge transfer from S–&N–heterocyclic inhibitors of BTA, 2MBT, 8HQ, and ATR as the electron donors adsorbed onto Cr–(Al–Ga) nanoalloy which play a role as the electron acceptor S $\rightarrow$ Cr, N $\rightarrow$ Cr. In fact, Cr sites in Cr–(Al–Ga) nanoalloy has more interaction energy from Van der Waals' forces with BTA, 2MBT, 8HQ, and ATR that can cause more stability towards coating specifications on the alloy crystal. It has been estimated that the precedence for picking out the crystal binding of N–, O–, S– atoms in BTA, 2MBT, 8HQ, and ATR in adsorption positions can be impressed by the essence of neighboring elements of aluminum and gallium in the Cr–(Al–Ga) surface.

3.4. *Infrared IR spectroscopy and thermodynamic analysis*

The IR spectra for adsorption of BTA, 2MBT, 8HQ, and ATR on the surfaces of Cr-doped on the Al–Ga have been reported in Figs. 5(a)–5(d). The graphs of Figs. 5(a)–5(d) have been observed in the frequency range between 500 cm^{-1} and 4500 cm^{-1} for the complexes of BTA @ Cr–(Al–Ga), 2MBT @ Cr–(Al–Ga), 8HQ @ Cr–(Al–Ga), and ATR @ Cr–(Al–Ga).

The IR spectrum for each of these materials in Figs. 5(a)–5(d) has been observed in the maximum frequency approximately around 500–4000 cm^{-1} with a sharp peak at 3500 cm^{-1} for BTA @ Cr–(Al–Ga) [Fig. 5(a)], around 500–3500 cm^{-1} with three sharps peak at 2000 cm^{-1}, 2100 cm^{-1} , and 2150 cm^{-1} for 2MBT @ Cr –(Al– Ga) [Fig. 5(b)], around 1000–3500 cm^{-1} with three sharp peaks at 1650 cm^{-1}, 1700 cm^{-1}, and 2000 cm^{-1} for 8HQ @ Cr–(Al–Ga) [Fig. 5(c)], around 1000–4500 cm^{-1} with two sharp peaks at 2000 cm^{-1} and 4325 cm^{-1} for ATR @ Cr–(Al–Ga) [Fig. 5(d)].

Furthermore, these materials have been accounted at CAM-B3LYP level of theory accompanying 6–31+G (d,p)/EPRIII/ LANL2DZ basis sets to receive the more valid equilibrium

Molecular simulation of (Al–Ga) surface 139

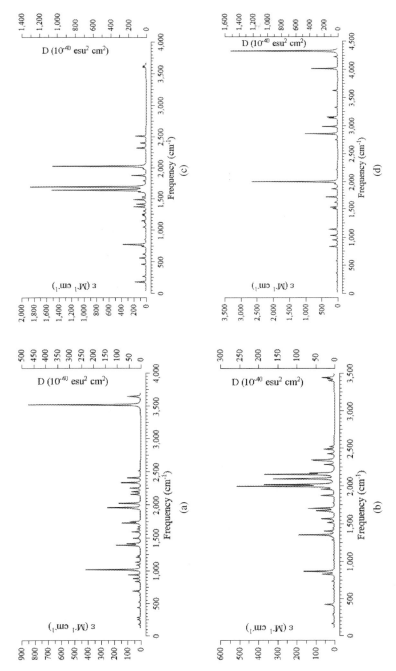

Fig. 5. Infrared spectra for chromium doping of aluminum-gallium nanosheet alloy as: (a) BTA @ Cr–(Al–Ga), (b) 2MBT @Cr–(Al–Ga), (c) 8HQ @Cr–(Al–Ga), and (d) ATR @ Cr–(Al–Ga).

Table 1. The physicochemical behavior of BTA, 2MBT, 8HQ, and ATR adsorbed on the Cr–(Al–Ga) surface and formation of BTA @ Cr–(Al–Ga), 2MBT @ Cr–(Al–Ga), 8HQ @ Cr–(Al–Ga), and ATR @ Cr–(Al–Ga) complexes.

Compound	$\Delta E° \times 10^{-4}$ (kcal/mol)	$\Delta H° \times 10^{-4}$ (kcal/mol)	$\Delta G° \times 10^{-4}$ (kcal/mol)	$S°$ (cal/K. mol)	Dipole momen (Debye)
Cr–Al–Ga	−751.2861	−751.2861	−751.2884	79.193	79.193
BTA	−24.5172	−24.5172	−24.5196	80.255	3.3166
2MBT	−69.5235	−69.5234	−69.5260	85.643	2.5108
8HQ	−29.5527	−29.5526	−29.5551	83.336	1.6389
ATR	−43.1336	−43.1336	−43.1358	73.839	1.5672
BTA@Cr–(Al–Ga)	−775.7408	−775.7408	−775.7435	91.111	4.9181
2MBT@Cr–(Al–Ga)	−820.7021	−820.7021	−820.7048	92.652	6.9180
8HQ@Cr–(Al–Ga)	−780.7942	−780.7941	−780.7970	96.292	5.8657
ATR@Cr–(Al–Ga)	−793.7160	−793.7159	−793.7187	92.282	6.2865

geometrical specifications, physical and thermodynamic parameter for each of the dedicated composition (Table 1).

Based on Table 1, the thermodynamic specifications, the authors concluded that this protective film containing the (BTA, 2MBT, 8HQ, and ATR)@Cr–(Al–Ga) which might be effective through doping of transition metal (Cr) toward formation of complexes including BTA @ Cr–(Al–Ga), 2MBT @ Cr–(Al–Ga), 8HQ @ Cr–(Al–Ga), and ATR @ Cr–(Al–Ga).

Considering Fig. 6, it could be detected that the maximum of the Langmuir adsorption isotherm curves on the basis of ΔG_{ads}^{o} may depend on the interactions between the BTA, 2MBT, 8HQ, and ATR heterocycles and the Cr–(Al–Ga) alloy. Comparing the ΔG_{ads}^{o} amounts versus dipole moment can affirm a good accord among computed consequences and the validity of the picked isotherm

Molecular simulation of (Al–Ga) surface 141

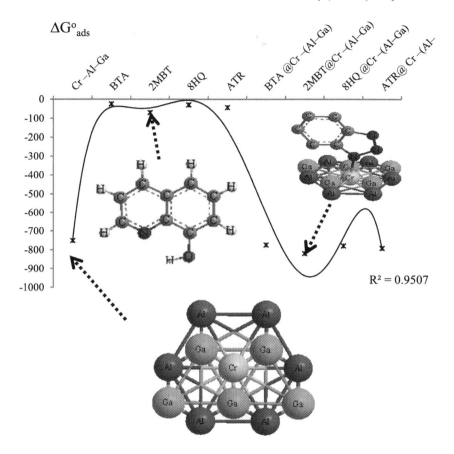

Fig. 6. Gibbs free energy for adsorbing process of BTA, 2MBT, 8HQ, and ATR on Cr–(Al–Ga) alloy surface.

for the adsorption procedure of BTA @ Cr–(Al–Ga), 2MBT @ Cr–(Al–Ga), 8HQ @ Cr–(Al–Ga), and ATR @ Cr–(Al–Ga) complexes (Fig. 6).

The adsorptive capacity of BTA, 2MBT, 8HQ, and ATR on the Cr–(Al–Ga) surface is affirmed by the $\Delta G°_{ads}$ quantities:

$$\Delta G°_{ads} = \Delta G°_{adsorbate@Cr\ (Al\text{-}Ga)} - (\Delta G°_{adsorbate} + \Delta G°_{Cr\ (Al\text{-}Ga)}). \quad (5)$$

Remarking Table 1, it is discovered that the adsorbing process of the S-&N–heterocyclic carbenes on the Cr–(Al–Ga) alloy might be physical and chemical nature. As seen in Table 1, all the accounted ΔG^{o}_{ads} amounts are very close, which demonstrate the agreement of the measured specifications by all methodologies and the reliability of the computing values.

Therefore, the interval between nitrogen, oxygen, sulfur atoms in BTA, 2MBT, 8HQ, and ATR during interaction and transition metal of Cr in Cr–(Al–Ga) nanoalloy, (N → Cr, O → Cr, S → Cr), has been estimated with relation coefficient of $R^2 = 0.9507$ (Table 1 and Fig. 6).

4. Conclusion

This study has analyzed the molecular properties of Al–Ga and the difference in its properties when doped with Transition metal Cr. In this work, the adsorbing process and penetration of BTA, 2MBT, 8HQ, and ATR on the Cr–doped Al–Ga crystals have been studied based on Langmuir theory using "ONIOM" methodology with ⟨⟨*high, medium* and *low*⟩⟩ levels of "EPR-III/6–31+G(d,p)/LANL2DZ, semi-empirical" and MM2 force fields accompanying Gaussian 16 revision C.01.

Denoting this research, the efficiency of the S–&N–heterocyclic carbenes" as the inhibiting agents for Cr–(Al–Ga) has been investigated through the electromagnetic and thermochemical traits extracts from PDOS, NMR, NQR, IR analysis which have been performed on BTA @ Cr–(Al–Ga), 2MBT @ Cr–(Al–Ga), 8HQ @ Cr–(Al–Ga), and ATR @ Cr–(Al–Ga) complexes.

Regarding NMR results, it can be observed the maximum fluctuation for gallium atoms in the neighboring of chromium in the center of metal alloy (adsorbent) for receiving the electrons of nitrogen (N → Cr), oxygen (O → Cr), and sulfur (S → Cr) of BTA, 2MBT, 8HQ, and ATR heterocycles (adsorbate).

In the preferred path, these S–&N–heterocyclic carbenes of BTA, 2MBT, 8HQ, and ATR stay collateral to the plane so long as fulfilling small single rotational steps with a C–C double bond belonged above a single transition metal element in Cr–(Al–Ga) nanoalloy. The inhibiting process of the protonated BTA, 2MBT, 8HQ, and ATR compositions is quoted to the total of the pure charge and the π charge of the six-ring. BTA, 2MBT, 8HQ, and ATR have been adsorbed on the crystal surface of the Cr–(Al–Ga) electrodes primarily in their protonated shapes. The most eventual adsorption pattern of the inhibiting materials is one in which the S &N–atoms of the BTA, 2MBT, 8HQ, and ATR are close to the Cr-doped of Al-Ga nanoalloy surface in an inclined state. The excellent molecule sensing performance is attributed to increased atom vacancies after doping and the increased adsorption of the molecules. This work presents an approach to synthesize other uniform metal-doped Al–Ga nanosurface and is also believed to be further extended to prepare other doped metal surface semiconductor nanostructures.

Acknowledgments

In successfully completing this paper and its research, the authors are grateful to Kastamonu University for their support through the office, library, and scientific websites.

Conflict of interest

The authors declare that they have no conflict of interest.

ORCID

Fatemeh Mollaamin ◎ https://orcid.org/0000-0002-6896-336X
Majid Monajjemi ◎ https://orcid.org/0000-0002-6665-837X

References

1. F. Mollaamin, A. Ilkhani, N. Sakhaei *et al.*, *J. Comput. Theor. Nanosci.* **12**, 3148 (2015). doi.org/10.1166/jctn.2015.4092.
2. C. Wei, *Technol. Innov. Appl.* **18**, 80 (2019).

3. H. M. Yang, Z. X. Qiu and G. Zhang, *Low Temperature Aluminum Electrolysis* (Northeast University Press, Shenyang, China, 2009).
4. H. M. Lu and Z. X. Qiu, *Light Metals* **4**, 24 (1997).
5. E. Machnikova, K. H. Whitmire and N. Hackerman, *Electrochim. Acta* **53**, 6024 (2008). doi: 10.1016/j.electacta.2008.03.021.
6. F. Mollaamin and M. Monajjemi, *J. Biol. Tribol. Corros.* **9**, 47 (2023). doi.org/10.1007/s40735-023-00768-3.
7. A. Fiala, A. Chibani, A. Darchen *et al.*, *Appl. Surf. Sci.* **253**, 9347 (2007). doi: 10.1016/j.apsusc.2007.05.066.
8. R. A. Prabhu, A. V. Shanbhag and T. V. Venkatesha, *J. Appl. Electrochem.* **37**, 491 (2007). DOI: 10.1007/s10800-006-9280-2.
9. G. C. Tian, J. Li and Y. X. Hua, *Chin. J. Process Eng.* **9**, 200 (2009).
10. G. C. Tian, J. Li and Y. X. Hua, *Trans. Nonferrous Met. Soc. China* **20**, 513 (2010).
11. M. Monajjemi, F. Mollaamin and S. Shojaei, *Biointerface Res. Appl. Chem.* **3**, 5575 (2020). doi.org/10.33263/BRIAC103.575585.
12. G. C. Tian, *Mater. Res. Found.* **54**, 249 (2019).
13. X. W. Zhong, T. Xiong, J. Lu and Z. N. Shi, *Nonferrous Met. Sci. Eng.* **5**, 44 (2014).
14. Y. Zheng, Q. Wang, Y. J. Zheng and H. C. Lv, *Chin. J. Process Eng.* **15**, 713 (2015).
15. V. Fleury, J. H. Kaufman and D. B. Hibbert, *Nature* **367**, 435 (1994).
16. G. Yue, X. Lu, Y. Zhu, X. Zhang and S. Zhang, *Chem. Eng. J.* **147**, 79 (2009).
17. F. Mollaamin and M. Monajjemi, **9**, 20 (2023). doi.org/10.3390/c9010020.
18. D. Esken, S. Turner, O. I. Lebedev, G. van Tendeloo and R. A. Fischer, *Chem. Mater.* **22**, 6393 (2010).
19. F. Mollaamin and M. Monajjemi, *Sens. Rev.* **43**, 266 (2023). doi.org/10.1108/SR-03-2023-0040.
20. B. Liu and B. Smit, *J. Phys. Chem. C* **114**, 8515 (2010).
21. S. Keskin, *J. Phys. Chem. C* **115**, 800 (2010).
22. U. P. N. Tran, K. K. A. Le and N. T. S. Phan, *ACS Catal.* **1**, 120 (2011).
23. J. VandeVondele, M. Krack, F. Mohamed, M. Parrinello, T. Chassaing and J. Hutter, *Comp. Phys. Comm.* **2**, 103 (2005).
24. A. Phan, C. J. Doonan, F. J. Uribe-Romo, C. B. Knobler, M. O'keeffe and O. M. Yaghi, *Acc. Chem. Res.* **43**, 58 (2010).
25. P. Hohenberg and W. Kohn, *Phys. Rev. B* **136**, B864 (1964).

26. W. Kohn and L. J. Sham, *Phys. Rev.* **140**, A1133 (1965).
27. F. Mollaamin and M. Monajjemi, *Clean Technol.* **5**, 403 (2023). doi. org/10.3390/cleantechnol5010020.
28. C. Hartwigsen, S. Goedecker and J. Hutter, *Phys. Rev. B* **58**, 3641 (1998).
29. J. P. Perdew, K. Burke and M. Ernzerhof, *Phys. Rev. Lett.* **77**, 3865 (1996).
30. J. VandeVondele and J. Hutter, *J. Chem. Phys.* **127**, 114105-1 (2007).
31. F. Mollaamin and M. Monajjemi, **11**, 108 (2023). https://doi. org/10.3390/computation11060108.
32. M. Mavrikakis, P. Stoltze and J. K. Nørskov, *Catal. Lett.* **64**, 101 (2000).
33. A. Dal Corso, *J. Phys.; Condens. Matter* **25**, 145401 (2013).
34. A. Tahan, F. Mollaamin and M. Monajjemi, *Russian J. Phys. Chem. A* **83**, 587 (2009).
35. F. Mollaamin and M. Monajjemi, *J. Mol. Model.* **29**, 170 (2023). https://doi.org/10.1007/s00894-023-05567-8.
36. H. Yildirim, T. Greber and A. Kara, *J. Phys. Chem. C* **117**, 20572 (2013).
37. M. Monajjemi, M. T. Baie and F. Mollaamin, *Russian Chem. Bull.* **59**, 886 (2010). https://doi.org/10.1007/s11172-010-0181-5.
38. M. Hoefling, F. Iori, S. Corni and K. E. Gottschalk, *ChemPhysChem.* **11**, 1763 (2010).
39. K. Bakhshi, F. Mollaamin and M. Monajjemi, *J. Comput. Theor. Nanosci.* **8**, 763 (2011). https://doi.org/10.1166/jctn.2011.1750.
40. H. Valencia, M. Kohyama, S. Tanaka and H. Matsumoto, *J. Chem. Phys.* **131**, 244705 (2009).
41. J. Clarke-Hannaford, M. Breedon, A. S. Best and M. J. Spencer, *Phys. Chem. Chem. Phys.* **21**, 10028 (2019).
42. Q. Q. Zhang, *Study on Electrodeposition of Aluminum and Aluminum Alloy in Ionic Liquid* (University of Chinese Academy of Sciences: Beijing, China, 2014).
43. F. Mollaamin and M. Monajjemi, *J. Mol. Model.* **29**, 119 (2023). doi. org/10.1007/s00894-023-05526-3.
44. S. A. Ali, M. A. J. Mazumder, M. K. Nazal and H. A. Al-Muallem, *Arab. J. Chem.* **13**, 242 (2020). doi: 10.1016/j.arabjc.2017.04.005.
45. H. Amar, J. Benzakour, A. Derja, D. Villemin and B. Moreau, *J. Electroanal. Chem.* **558**, 131 (2003). doi: 10.1016/S0022-0728(03)00388-7.
46. M. S. El-Sayed, *Int. J. Electrochem. Sci.* **6**, 1479 (2011).

47. E. S. Sherif, *Molecules* **19**, 9962 (2014). doi: 10.3390/molecules19079962).
48. D. Yang, M. Zhang, J. Zheng and H. Castaneda, *RSC Adv.* **5**, 160 (2015). doi: 10.1039/C5RA14556B.
49. M. Finšgar and I. Milošev, *Corros. Sci.* **52**, 2737 (2010). Doi:10.1016/j.corsci.2010.05.002.
50. A. Y. Shen, S. N. Wu and C. T. Chiu, *J. Pharm. Pharmacol.* **51**, 543 (1999). doi:10.1211/0022357991772826.
51. P. Cabassi *et al.*, *J. South African Inst. Mining Metall.* **83**, 270 (1983).
52. M. Monajjemi, F. Mollaamin, M. R. Gholami *et al.*, *Met. Chem.* **26**, 349 (2003). https://doi.org/10.1515/MGMC.2003.26.6.349.
53. F. Mollaamin, S. Shahriari, M. Monajjemi *et al.*, *J. Clust. Sci.* **34**, 1547 (2023). https://doi.org/10.1007/s10876-022-02335-1.
54. F. Mollaamin and M. Monajjemi, *J. Biol. Tribol. Corros,* **9**, 33 (2023). https://doi.org/10.1007/s40735-023-00751-y.
55. F. Mollaamin and M. Monajjemi, *Mol. Simul.* **49**, 365 (2023). https://doi.org/10.1080/08927022.2022.2159996.
56. F. Mollaamin and M. Monajjemi, *J. Clust. Sci.* (2023). https://doi.org/10.1007/s10876-023-02436-5.
57. A. A. Bukaemskii, A. G. Beloshapko and A. P. Puzyr', *Fiz. Goreniya Vzryva* **36**, 119 (2000).
58. A. A. Bukaemskii, L. S. Tarasova and E. N. Fedorova, *Izv. Vyssh. Uchebn. Zaved., Tsvetn. Metall.,* **5**, 60 (2000).
59. Im Tkhek-de, N. E. Lyamkina, A. L. Lyamkin *et al.*, *Pis'ma Zh. Tekh. Fiz.* **27**, 10 (2001).
60. N. E. Lyamkina, G. A. Chiganova, V. V. Slabko *et al.*, *Inorg. Mater.* **41**, 830 (2005). https://doi.org/10.1007/s10789-005-0221-y.
61. P. Görlikh, H. Karras, H. Kötits and R. Leman, Spektroskopicheskie svoistva aktivirovannykh lazernykh kristallov (Spectroscopic Properties of Activated Laser Crystals), Moscow: Nauka, (1966).
62. F. Zupanič, S. Žist, M. Albu, I. Letofsky-Papst, J. Burja, M. Vončina and T. Bončina, *Materials* **16**, 2949 (2023). https://doi.org/10.3390/ma16082949.
63. S. Li, H. Dong, X. Wang and Z. Liu, *Materials* **12**, 1595 (2019). https://doi.org/10.3390/ma12101595.
64. Z. Wang, P. Zhang, X. Zhao and S. Rao, *Coatings* **12**, 1899 (2022). https://doi.org/10.3390/coatings12121899.

65. B. Ghalandari, M. Monajjemi and F. Mollaamin, *J. Comput. Theor. Nanosci.* **8**, 1212 (2011). doi.org/10.1166/jctn.2011.1801.

66. M. Svensson, S. Humbel, R. D. J. Froese, T. Matsubara, S. Sieber and K. Morokuma, *J. Phys. Chem.* **100**, 19357 (1996). https://doi.org/10.1021/jp962071j.

67. F. Mollaamin and M. Monajjemi, *Russ. J. Phys. Chem. B* **17**, 658 (2023). https://doi.org/10.1134/S1990793123030223.

68. F. Brandt and C. R. Jacob, *J. Chem. Theor. Comput.* **18**, 2584 (2022). https://doi.org/10.1021/acs.jctc.1c01093.

69. P. E. Blöchl, *Phys. Rev. B* **50**, 17953 (1994).

70. J. P. Perdew, K. Burke and M. Ernzerhof, *Phys. Rev. Lett.* **77**, 3865 (1996).

71. P. Ziesche, S. Kurth and J. P. Perdew, *Comput. Mater. Sci.* **11**, 122 (1998).

72. M. Arrigoni and G. K. H. Madsen, *Comput. Mater. Sci.* **156**, 354 (2019).

73. A. D. Becke, *J. Chem. Phys.* **98**, 5648 (1993).

74. C. Lee, W. Yang and R. G. Parr, *Phys. Rev. B* **37**, 785 (1988).

75. K. Kim and K. D. Jordan, *J. Phys. Chem.* **98**, 10089 (1994). Doi:10.1021/j100091a024.

76. F. Mollaamin, M. Monajjemi, S. Salemi and M. T. A. Baei, *Fuller. Nanotub. Carbon Nanostruct.* **19**, 182 (2011). doi.org/10.1080/15363831003782932.

77. C. J. Cramer, *Essentials of Computational Chemistry: Theories and Models*, 2nd edn. (Wiley, 2004). Wiley.com. Retrieved 2021-06-24.

78. M. Khaleghian, M. Zahmatkesh, F. Mollaamin and M. Monajjemi, *Fuller. Nanotub. Carbon Nanostruct.* **19**, 251 (2011). doi.org/10.1080/15363831003721757.

79. M. J. Frisch, G. W. Trucks, H. B. Schlegel, G. E. Scuseria, M. A. Robb, J. R. Cheeseman, G. Scalmani, V. Barone, G. A. Petersson, H. Nakatsuji, X. Li, M. Caricato, A. V. Marenich, J. Bloino, B. G. Janesko, R. Gomperts, B. Mennucci, H. P. Hratchian, J. V. Ortiz, A. F. Izmaylov, J. L. Sonnenberg, D. Williams-Young, F. Ding, F. Lipparini, F. Egidi, J. Goings, B. Peng, A. Petrone, T. Henderson, D. Ranasinghe, V. G. Zakrzewski, J. Gao, N. Rega, G. Zheng, W. Liang, M. Hada, M. Ehara, K. Toyota, R. Fukuda, J. Hasegawa, M. Ishida, T. Nakajima, Y. Honda, O. Kitao, H. Nakai, T. Vreven, K. Throssell, J. A. Jr. Montgomery, J. E. Peralta, F. Ogliaro, M. J. Bearpark,

J. J. Heyd, E. N. Brothers, K. N. Kudin, V. N. Staroverov, T. A. Keith, R. Kobayashi, J. Normand, K. Raghavachari, A. P. Rendell, J. C. Burant, S. S. Iyengar, J. Tomasi, M. Cossi, J. M. Millam, M. Klene, C. Adamo, R. Cammi, J. W. Ochterski, R. L. Martin, K. Morokuma, O. Farkas, J. B. Foresman and D. J. Fox, Gaussian 16, Revision C.01, Gaussian, Inc., Wallingford CT (2016).

80. GaussView, Version 6.06.16, Dennington, Roy; Keith, Todd A.; Millam, John M. Semichem Inc., Shawnee Mission, KS (2016).

81. U. Sohail, F. Ullah, N. H. Binti Zainal Arfan, M. H. S. Abdul Hamid, T. Mahmood, N. S. Sheikh and K. Ayub, *Molecules* **28**, 4060 (2023). https://doi.org/10.3390/molecules28104060.

82. M. Monajjemi, M. Khaleghian, N. Tadayonpour and F. Mollaamin, *Int. J. Nanosci.* **9**, 517 (2010). doi.org/10.1142/S0219581X10007071.

83. M. A. A. Zadeh, H. Lari, L. Kharghanian, E. Balali, R. Khadivi, H. Yahyaei, F. Mollaamin and M. Monajjemi, *J. Comput. Theor. Nanosci.* **12**, 4358 (2015).

84. E. M. Sarasia, S. Afsharnezhad, B. Honarparvar, F. Mollaamin and M. Monajjemi, *Phys. Chem. Liq.* **49**, 561 (2011). doi.org/10.1080/00319101003698992.

85. A. Young, Hugh and R. D. Freedman, *Sears and Zemansky's University Physics with Modern Physics*, 13th edn. (Addison-Wesley, Boston, 2012), p. 754.

Chapter 9

Photothermal insulation mechanism of submicron ITO hollow particles in PVDF film

Hedong Li ⊙*, Peihu Shen ⊙*, Zizheng He ⊙[†,‡], Yang Xu ⊙[†]
and Minjia Wang ⊙[*,§]

*School of Civil Engineering and Architecture
Zhejiang Sci-Tech University, Hangzhou 310018, P. R. China
[†]College of Materials and Chemistry, China Jiliang University
Hangzhou 310018, P. R. China
[‡]StarPower Microelectronics Co, Ltd
Jiaxing 314000, P. R. China
[§]coolwangmin@zstu.edu.cn

In this study, submicron Sn^{4+}-doped In_2O_3 hollow particles (indium–tin oxide (ITO)) with a lyrium shell-like surface were synthesized via solvothermal method. The appropriate addition of these particles into polyvinylidene fluoride (PVDF) films exhibits high visible light transmission and effective UV/IR blocking properties, surpassing those of ITO nanoparticles. This can be attributed to the reduced absorption of UV/IR radiation by the hollow ITO particles, as well as their higher diffuse reflectivity and thermal insulation.

Keywords: ITO hollow particles; film; photothermal barrier.

[§]Corresponding author.
To cite this article, please refer to its earlier version published in the Functional Materials Letters, Volume 16(7), 2340029 (2023), DOI: 10.1142/S1793604723400295.

1. Introduction

Building energy consumption constitutes nearly 40% of the total social energy consumption, and building energy-saving materials are crucial in the face of increasing demand for energy.[1-3] Glass film is a highly efficient building energy-saving material that can significantly strengthen the photothermal barrier properties of ordinary glass. The material, with functional fillers as its core materials,[4-6] mainly including SnO_2,[7] TiO_2,[8] AZO,[9] ATO,[10] ITO[11] and other transparent conductive oxides (TCO), generally exhibits high transmission in the visible region, UV cut-off characteristics, and high reflection in the IR region. Among them, indium-tin oxide (ITO), as an n-type semiconductor with high transparency, low resistivity, and wide bandgap,[12] is widely used in liquid crystal displays,[13] solar cells,[14] and gas-sensitive sensors,[15] etc. The morphological structure of ITO particles can impact various properties. For instance, ITO nanoflowers can enhance their dispersion in polyurethane acrylate (PUA),[16] and also improve the resistivity of thin films.[17] ITO nanorods can effectively improve the optoelectronic properties and hydrophilicity of thin films.[18] To date, the research on ITO materials mainly focuses on the optoelectronic properties of thin films, with limited investigation into the underlying mechanisms responsible for their enhancement in thermal insulation properties. In this study, Sn^{4+} doped In_2O_3 hollow particles were prepared using a template-free method and compounded in polyvinylidene fluoride (PVDF) films to achieve high visible transmission as well as effective barrier properties of UV and IR radiation.

2. Experimental

2.1. *Materials*

The raw materials utilized were Stannous chloride pentahydrate (99%, Aladdin), indium chloride (99%, Maclean), anhydrous ethanol (AR, Maclean), polyethylene glycol 400 (AR, Maclean), ethylene glycol (98%, Maclean), n-butanol (99%, Maclean), poly vinylidene fluoride (AR, Shandong Dongyue Shenzhou New

Material Co., Ltd.), and commercial nano-ITO (industrial pure, Bisili New Material Co., Ltd.).

2.2. *Preparation of ITO-PVDF glass films*

Herein, 0.18 mmol indium chloride was dissolved in a mixture of 50 mL anhydrous ethanol, 10 mL polyethylene glycol, and 10 mL ethylene glycol. Subsequently, 10 mL n-butanol was added dropwise and stirred well. Then 0.02 mmol tin tetrachloride pentahydrate was added into the reaction system, which underwent solvothermal treatment at 180°C for 6 h to yield a white precipitate that was obtained through washing, centrifuging, and drying. Afterward, the ITO hollow particles were obtained through heat treatment at 500°C for 2 h. Then, 1 g PVDF powder was dissolved in 12 mL DMF solution. And varying amounts of ITO hollow particles were added and stirred for 48 h until uniformly dispersed in the film. The resulting film was spin-coated on the surface of ordinary glass using a spin-coating machine and dried to completely evaporate the DMF solvent, yielding a series of ITO-PVDF glass films.

2.3. *Characterization*

The hydrothermal reaction products were analyzed using a simultaneous thermal analyzer (DSC, PE, Netzsch, Germany) through thermogravimetric-differential scanning calorimetry from 30°C to 950°C. The microscopic morphology of the hollow particles was characterized by field emission scanning electron microscopy (FESEM, Zeiss Sigma 300, Germany). Additionally, the lattice structure of the hollow particles was observed by high-resolution transmission electron microscopy (HRTEM, JEM-1200EX, JEOL). The absorption and diffuse reflection spectra of the particles, as well as the transmittance spectra of the films, were characterized by UV-VIS-IR spectrophotometer (Shimadzu, UV-3600). The photothermal barrier performance of the coated glass was tested using a custom-built thermal insulation chamber, as depicted in Fig. 1. The chamber was divided into two compartments with the coated glass placed in the

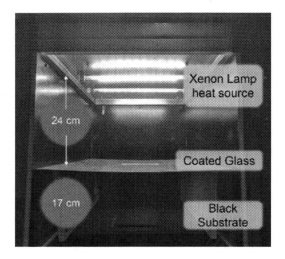

Fig. 1. Internal structure of the custom-built thermal insulation chamber.

center. Four bath lamps, totaling 800 W and positioned 24 cm away from the coated glass and 17 cm away from the black polystyrene foam substrate, served as the photothermal source. A thermocouple thermometer with an accuracy of 0.1°C and an error margin of 0.2°C can be employed to measure the temperature of both the coated glass backplane and polystyrene foam substrate surface, enabling assessment of the thermal barrier performance.

3. Results and Discussion

3.1. *Phase structure and morphology*

Figure 2 shows the XRD pattern of the white precipitate prepared by the solvothermal method. The spectral lines exhibit a broad peak pattern, indicating that the white precipitate is amorphous and likely to be indium-tin hydroxide.[19] The TG-DSC results depicted in Fig. 3 reveal a significant decrease for the white precipitate within the temperature range of 25–390°C. A weight loss of 13% was observed below 200°C, which can be attributed to the volatilization of adsorbed water on the sample surface and residual organic matter such as ethylene glycol. The weight loss of 20.6%

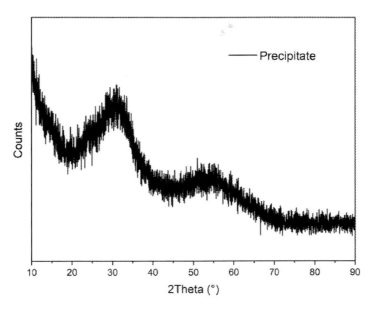

Fig. 2. XRD patterns of indium-tin hydroxide.

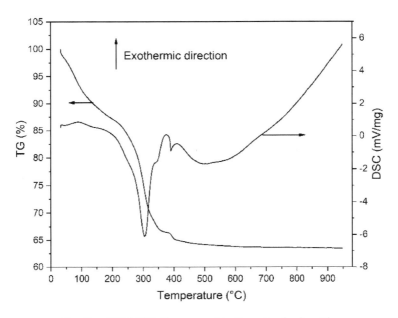

Fig. 3. TG-DSC diagram of indium-tin hydroxide.

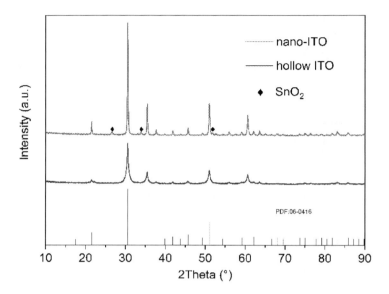

Fig. 4. XRD patterns of ITO hollow particles and ITO nanoparticles.

observed at temperatures ranging from 200°C to 390°C, along with a heat absorption peak near 300°C in the DSC curve, can be attributed to the dehydration process of indium-tin hydroxide.

Figure 4 displays the XRD patterns of hollow ITO and nano-ITO. The former exhibits a cubic ferromanganese ore structure (PDF:06–0416) with a space group of Ia-3(206), devoid of any impurity phase, indicating that Sn^{4+} enters the In_2O_3 lattice in the form of dopant ions. The latter contain not only the cubic ferromanganese ore structure of In_2O_3 but also an impurity phase SnO_2 (PDF:41–1445).

Figure 5 displays the SEM images of ITO hollow particles and ITO nanoparticles. Different from the nanoparticles depicted in Fig. 5(c), the hollow particles exhibit a spherical shape and particle size ranging from 350 nm to 550 nm. The surface morphology of the hollow particles resembles that of a lyrium shell, as shown in Fig. 5(d).

Figure 6 displays the TEM of the hollow ITO. The bright-field image clearly reveals the hollow structure from the mass-thickness liner, while high-resolution mapping confirms that the crystal plane spacing of (222) is 0.288 nm, consistent with the XRD results.

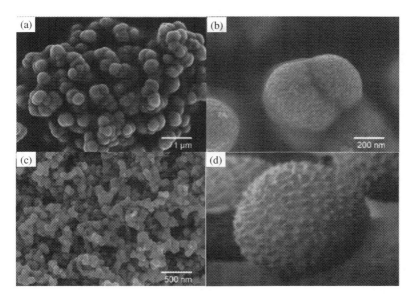

Fig. 5. SEM images of ITO hollow particles at (a) low magnification and (b) high magnification, (c) ITO nanoparticles, (d) Lychee Image.

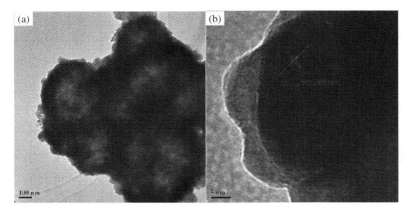

Fig. 6. (a) TEM and (b) HRTEM patterns of ITO hollow particles.

Figure 7(a) shows the UV-Vis-IR absorption spectra of ITO hollow particles and ITO nanoparticles. The absorbance exhibits a non-monotonic dependence on wavelength, with a minimum in the visible region and distinct absorption features in both the UV and

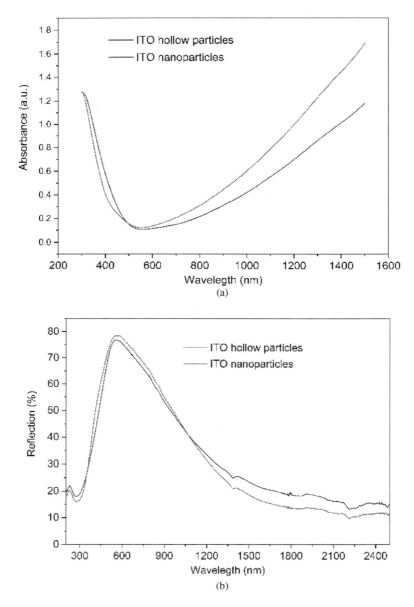

Fig. 7. UV-Vis-IR (a) absorption and (b) diffuse reflection spectra of ITO hollow particles and ITO nanoparticles.

IR regions. Specifically, the UV absorption can be attributed to high-energy transitions resulting from intrinsic ITO properties, while broad IR absorption arises from surface plasma resonance of ITO.[20] The weaker absorption of hollow ITO in the IR region compared to nano-ITO is attributed to the larger specific surface area of nano-ITO. Figure 7(b) displays the diffuse reflectance spectra of both particles, revealing that the lyrium shell-like rough surface of hollow ITO results in higher diffuse reflectance in both UV and IR regions.

3.2. Analysis of photothermal insulation

Figure 8 displays the UV-Vis-IR transmittance spectra of various ITO-PVDF films. The addition of these particles results in a photothermal blocking effect, causing a gradual decrease in

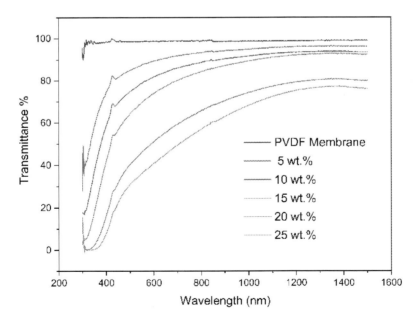

Fig. 8. UV-Vis-IR transmittance spectra of ITO-PVDF films with varying levels of addition.

transmittance as the amount increased. Additionally, ITO primarily absorbs blue-violet and red light. The average visible transmittance of the ITO-PVDF films can be calculated from the following equation[21]:

$$T_{vis} = \frac{\int_{400}^{780} T(\lambda)i(\lambda)\,d\lambda}{\int_{400}^{780} i(\lambda)\,d\lambda} \times 100\% \approx \frac{\sum_{\lambda=400}^{780} T(\lambda)i(\lambda)}{\sum_{\lambda=400}^{780} i(\lambda)} \times 100\%, \quad (1)$$

where T_{vis} is the visible transmittance, $T(\lambda)$ is the spectrophotometer test transmittance data, and $i(\lambda)$ is the irradiation intensity of light at wavelength λ. The basic parameter is sourced from ASTM G173 solar irradiation intensity and the calculation results are shown in Table 1. The average visible transmittance of the ITO-PVDF films remains above 70% when the addition amount is below 15 wt.%. When the addition amount reaches 20 wt.%, there is a rapid decrease of over 20% in the visible light transmittance for the coated glass. Upon further addition to 25 wt.%, the average visible light transmittance drops to only 40.33%.

The temperature change curve of the coated glass backplane and the substrate in the thermal insulation chamber is depicted in Fig. 9. Figure 9(a) illustrates the temperature curve of the coated glass backplane, where it can be observed that the pure glass

Table 1. The average transmittance of the ITO-PVDF films.

Samples	Vis transmittance (%)
0 wt.%	98.76
5 wt.%	88.62
10 wt.%	80.63
15 wt.%	72.41
20 wt.%	50.08
25 wt.%	40.33

Photothermal insulation mechanism of submicron   159

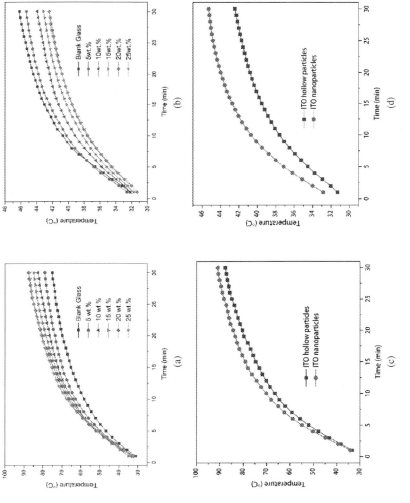

Fig. 9. Temperature rise curves of (a) and (c) coated glass backplane and (b) and (d) substrate surfaces with different additions.

exhibits a slower temperature rise rate compared to samples with arbitrary addition due to the absorption of ITO particles. Figure 9(b) depicts the temperature profile of the substrate, wherein an increase in hollow particles results in a deceleration of the temperature rise rate. The sample with 20 wt.% addition exhibits the highest backplane temperature and the lowest substrate temperature, measuring 87.69°C and 42.36°C on the coated glass backplane and substrate surface, respectively, after 30 min of irradiation.

When the addition amount reaches 25 wt.%, the substrate temperature does not exhibit a continuous decrease, and the temperature of the glass backplane ceases to rise, indicating that a dynamic balance among absorption, reflection, and heat conduction has been achieved at 20 wt.%. A comparison of ITO hollow particles and ITO nanoparticles added at 20 wt.% in terms of their thermal insulation performance is presented in Figs. 9(c) and 9(d). The temperature rise rate of both the hollow particle-coated glass backplane and substrate is lower than that of the nanoparticle film, with approximately 3.2°C and 2.9°C lower temperature on the coated glass backplane and substrate surface, respectively, after 30 min of irradiation. This can be attributed to the reduced UV and IR light absorption by hollow ITO, higher diffuse reflectance compared to nanoparticles, as well as hindered heat transfer caused by its unique hollow structure.

4. Conclusions

In this study, ITO hollow particles were synthesized via a template-free method using solvent thermal combined with heat treatment technique and then compounded in PVDF films. The photothermal insulation properties of ITO-PVDF films were thoroughly examined and found to be exceptional, leading to the following conclusions:

(1) The particle surface showed a lyrium shell morphology and the particle size ranged from 350 nm to 550 nm. ITO-PVDF films exhibited excellent photothermal insulation property with the

addition amount of 20 wt.%. After 30-min irradiation, the coated glass backplane reached a temperature of 87.69°C, while the polystyrene foam substrate surface reached a temperature of 42.36°C.

(2) The ITO-PVDF films effectively blocked UV and IR radiation compared to ITO nanoparticles. The coated glass backplane and polystyrene foam substrate surface temperatures of hollow ITO were approximately 3.2°C and 2.9°C lower than those of nano-ITO, respectively, after 30-min irradiation at 20 wt.% addition. This can be attributed to the rough surface and hollow structure, resulting in less absorption of UV and IR light, higher diffuse reflectance, and impeded heat transfer.

Acknowledgments

The authors gratefully acknowledge the financial support from Natural Science Foundation of Zhejiang Province (No. LQ20E080019), China Postdoctoral Science Foundation (No. 2020M681920), Zhejiang postdoctoral merit-based funding (ZJ2020030), and Zhejiang Sci-Tech University Youth Innovation Program (2021Q038).

ORCID

Hedong Li ◉ https://orcid.org/0000-0002-0911-1976
Peihu Shen ◉ https://orcid.org/0009-0008-2394-0624
Zizheng He ◉ https://orcid.org/0009-0003-6352-1129
Yang Xu ◉ https://orcid.org/0000-0002-5027-4339
Minjia Wang ◉ https://orcid.org/0000-0003-2954-9999

References

1. J. Gao *et al.*, *Nat. Commun* **10**, 1 (2019).
2. Q. Gao, X. Wu and T. Huang, *Ceram Int* **47**, 23827 (2021).
3. S. D. Rezaei, S. Shannigrahi and S. Ramakrishna, *Sol. Energ. Mat. Sol. C* **159**, 26 (2017).
4. X. Xu *et al.*, *Sol. Energ Mat. Sol. C* **168**, 119 (2017).

5. S. Qi *et al.*, *Crystengcomm* **21**, 3264 (2019).
6. B. Shen *et al.*, *Ceram. Int.* **46**, 18518 (2020).
7. M. Anu and S. S. Pillai, *Solid State Commun* **341**, 114577 (2022).
8. H. Yu *et al.*, *Chem. Eng. J* **418**, 129342 (2021).
9. S. Soumya *et al.*, *Sol. Energ. Mat. Sol. C* **143**, 335 (2015).
10. M. Wang *et al.*, *Ceram. Int.* **47**, 547 (2021).
11. H. Matsui *et al.*, *ACS Appl. Nano Mater* **1**, 1853 (2018).
12. M. Ezzeldien *et al.*, *Mater. Today Commun.* **31**, 103272 (2022).
13. X. Yan *et al.*, *Jpn. J. Appl. Phys.* **48**, 120203 (2009).
14. J. H. Kim *et al.*, *J. Power Sources* **418**, 152 (2019).
15. A. S. Mokrushin *et al.*, *Talanta* **221**, 121455 (2021).
16. H. Liu *et al.*, *Appl. Surf. Sci,* **258**, 8564 (2012).
17. X. Wu, Y. Wang and B. Yang, *Appl. Phys. A-Mater.* **117**, 781 (2014).
18. H. K. Park *et al.*, *J. Mater. Chem. A* **1**, 5860 (2013).
19. X. Xu and X. Wang, *Inorg. Chem.* **48**, 3890 (2009).
20. Q. Gao, X. Wu and T. Huang, *Sol. Energy* **220**, 1 (2021).
21. Q. Gao, X. Wu and D. Wang, *Sol. Energy* **238**, 60 (2022).

Chapter 10

Effect of silica morphology on rheological properties and stability of magnetorheological fluid

Bingsan Chen ⊕*,†, Minghan Yang*, Chenglong Fan*, Xiaoyu Yan*, Yongchao Xu* and Chunyu Li*

Fujian Key Laboratory of Intelligent Machining Technology and Equipment Fujian 350118, P. R. China
†*bschen126@fjut.edu.cn*

SiO_2 is commonly used as an abrasive in optical devices and magnetor-heological (MR) fluid. In this study, MR fluid with composite and free-state SiO_2 were prepared. Composite magnetic particles of $CIP@SiO_2$ were synthesized using the sol–gel method with tetraethyl orthosilicate (TEOS) as the silicon source. Performance test results showed that $CIP@SiO_2$ MR fluid exhibited superior mechanical properties compared with CIP MR fluid under zero magnetic field conditions. However, under the influence of a magnetic field, the SiO_2 shell weakened the response of $CIP@SiO_2$ MR fluid to the magnetic field, whereas CIP MR fluid demonstrated better rheological properties. The addition of free-state SiO_2 to CIP MR fluid significantly improved its rheological performance. At a volume fraction of 6% SiO_2, the shear stress and viscosity reached their maximum values, and further increasing the volume fraction to 8% resulted in a noticeable decrease in rheological behavior. Stability improved with an increase in magnetizable particle

---

†Corresponding author.

To cite this article, please refer to its earlier version published in the Functional Materials Letters, Volume 16(7), 2340030 (2023), DOI: 10.1142/S1793604723400301.

content. The stability of CIP MR fluid at the same concentration was superior to that of CIP@SiO$_2$ MR fluid, while free-state SiO$_2$ had a notable enhancing effect on stability. Therefore, when SiO$_2$ exists as a free-state abrasive, the rheological properties and stability of the MR fluid considerable improve.

Keywords: Magnetorheological fluid; silica; composite magnetic particles; rheology; stability.

1. Introduction

Magnetorheological (MR) fluid is a type of intelligent material that consists of a carrier fluid, magnetizable particles, and additives. In the absence of a magnetic field, it behaves like a low-viscosity Newtonian fluid. However, when a specific magnetic field is applied, the magnetizable particles align themselves in chains along the direction of the magnetic field due to the influence of magnetic forces. This alignment causes the fluid to exhibit the properties of a Bingham fluid, resulting in a substantial increase in viscosity and yield strength.[1,2] MR fluid has diverse applications in various fields, including construction, medicine, and precision machining. The performance requirements of this fluid vary depending on the specific field of application. In the construction field, MR fluid is utilized in seismic dampers to mitigate the hazards caused by earthquakes. At this point, the MR fluid must exhibit significant damping force.[3] In the medical field, MR fluid is applied in prostheses, exoskeletons, and orthopedic devices, to enhance the stability of limb movement. Therefore, MR fluid must demonstrate improved stability performance.[4] Additionally, MR polishing is commonly employed in the processing of optical components. This technique offers advantages such as minimal surface roughness, high precision in surface type, minimal surface damage, and easy control of the processing process. Consequently, the MR fluid used in this context must possess enhanced rheological properties. The main MR polishing technologies in the field of precision machining include the following: cluster MR polishing technology, combined MR polishing technology, and ultrasonic composite MR polishing technology. The basic principle of these technologies involves adding micro and nano abrasive particles to the MR fluid to create small polishing

Effect of silica morphology on rheological properties 165

heads, which are formed under the influence of a magnetic field and are used to remove materials from the surface of the workpiece being polished. The goal is to achieve a high surface accuracy.[5] The rheological properties of MR fluid, such as shear yield stress, must be considered as they play a critical role in the process. A notable detail that when the density of magnetizable particles is high, issues such as agglomeration and sedimentation may occur, which can lead to failure.[6,7] Therefore, improving the stability of these fluids is necessary to extend their service life.

Currently, researchers often employ surfactants and thickeners as additives to enhance the performance of MR fluid. For example, Maurya *et al.* added nanoscale bentonite and oleic acid to MR fluid and demonstrated a transition from linear viscoelasticity to nonlinearity and a significant enhancement in storage modulus.[8] Bombard *et al.* investigated the influence of structurally similar non-magnetic goethite nanofibers and chromium dioxide nanofibers on the yield stress of MR fluid. Both types of nanofibers, at appropriate concentrations, increased the yield stress of the MR fluid, indicating that particle shape anisotropy is a key factor in performance improvement.[9] Milde *et al.* prepared MR polishing fluid by using ferromagnetic particles, Al_2O_3 and nanoscale clay in the form of halloysite, which greatly improved the sedimentation stability of the MR polishing fluid.[10] Silicon dioxide, as a common inorganic material, is a frequently used additive in MR fluid. Kaide *et al.* studied the effects of three additives, including styrene–isoprene–styrene block copolymers, hydrophobic fumed silica, and organic gelators (CMOL, SB03, PMDA–2C8/oleyl), on the performance of MR fluid. The addition of block copolymers effectively met the stability requirements of the MR fluid and these additives enhanced its magnetic properties.[11] Chen *et al.* investigated the influence of non-ferromagnetic particles on the performance of MR fluid and developed a viscosity calculation model considering the dynamic magnetic field and the temperature field. The experiments demonstrated that adding a certain amount of nanoglass powder improved the viscosity and stability of the MR fluid and enhanced its magnetic saturation intensity under a dynamic magnetic field.[12] Xu *et al.* studied the effect of mass fraction of hydrophobic fumed silica on the rheological and sedimentation properties of MR fluid. The results showed that adding hydrophobic fumed silica significantly increased the viscosity and yield stress of the MR fluid, whereas the shear thinning index decreased sharply. The optimal addition ratio was found to be 0.5–0.6 wt.%.[13] Bombard *et al.* investigated the effects of CIP, phosphated CIP, hydrophobic silica, and hydrophilic silica on MR fluids. They revealed that the types of silica and their specific interactions with the phosphated coating on iron powder drove the rheological behavior of the MR fluid in different shear rate regions in the absence of a magnetic field. In the presence of a magnetic field, the differences in magnetic powder and their magnetic properties primarily influenced the rheological curves, particularly at higher shear rates.[14] Aruna *et al.* studied the effects of hydrophobic and hydrophilic fumed silica mixed with silicone oil as additives on MR fluid. The results showed that hydrophilic silica with a larger surface area exhibited higher yield stress values and good sedimentation performance within 7 days.[15]

In summary, the existing research on the effects of silicon dioxide on MR fluid achieved significant progress, but it mainly focused on hydrophobic fumed silica and nanoscale silica. Further research on silicon dioxide materials that could effectively polish is needed to meet the requirements of MR polishing. In this study, composite magnetic polishing particles (CIP@SiO$_2$) were prepared using the sol–gel method, characterized and analyzed. Different forms of MR fluid, including carbonyl iron powder–based (CIP), CIP@SiO$_2$ and mixed particle–based (CIP/SiO$_2$) fluids, were prepared and tested to investigate the influence of different forms of silicon dioxide on the performance of MR fluid. The results provide valuable references for the preparation of MR fluid.

2. Preparation of Core–Shell Structure

In this experiment, the sol–gel method was adopted to prepare CIP@SiO$_2$ magnetizable particles. The reaction principle is the hydrolysis and condensation reaction of tetraethyl orthosilicate

(TEOS) under alkaline environment with the following reaction equation[16-18]:

$$(C_2H_5O)_4Si + 2H_2O \xrightarrow{OH^-} 4C_2H_5OH + SiO_2. \tag{1}$$

A certain amount of carbonyl iron powder was taken and washed with anhydrous ethanol to remove surface impurities. Then, the iron powder was fully dispersed using ultrasound and dried in a drying oven. The dried iron powder (9.8 g) was weighed and placed in a beaker, followed by the addition of 180 mL of anhydrous ethanol. Ultrasound dispersion was conducted for 20 min. Next, 40 mL deionized water, 6 mL ammonia water, and 12.48 g TEOS were added separately. The mixture was mechanically stirred at 35°C for 12 h with a stirring speed of 500 r/min. After the reaction, the iron powder was accelerated to settle by using a magnet, and the obtained iron powder was dried at 60°C for 24 h in an ambient environment for testing. The specific process is shown in Fig. 1.

3. Characterization Results and Discussion

The transmission electron microscope (TEM) images of carbonyl iron powder before and after coating are shown in Fig. 2. The scale bars indicate measurements of 1 μm, 500 nm, and 100 nm, respectively.

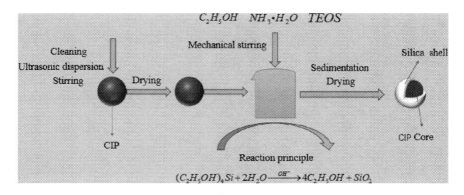

Fig. 1. Preparation flow chart.

168   B. Chen et al.

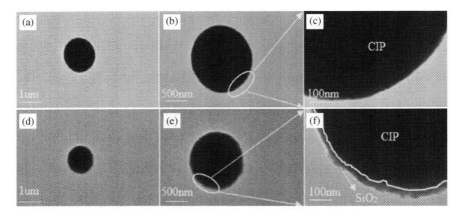

Fig. 2. TEM images of CIP(a),(b), (c)and CIP@SiO$_2$ (d),(e),(f).

In images (a), (b), and (c), the same CIP particle is shown at different magnifications. The surface of the CIP particle appeared to be relatively smooth and is not covered by any other substances. Meanwhile, images (d), (e), and (f) depict the CIP@SiO$_2$ particle, demonstrating a rougher surface in ima€(e). The presence of an obvious core–shell structure and uniform distribution of the external shell can be observed. This finding suggested that a new material was generated, uniformly covering the external part of the CIP particle, with the shell thickness measuring approximately 20 nm.

Field emission scanning electron microscopy (FE-SEM) images and energy-dispersive X-ray (EDX) elemental maps were utilized to analyze the composition of CIP@SiO$_2$ composite magnetizable particles, as depicted in Fig. 3. The elemental distribution of the composite particles revealed the presence of four main elements: Fe, C, O, and Si. Fe, C, and O are inherent elements of carbonyl iron powder, and Si represents the silicon dioxide coating on the particle surface. Figure 3(f) demonstrates the uniform distribution of silicon on the surface of the carbonyl iron powder microspheres, confirming the uniform coating of SiO$_2$ on CIP particles.

Figure 4(a) displays the XRD patterns of CIP and CIP@SiO$_2$ particles. Both samples exhibited three identical diffraction peaks at $2\theta = 44.7°$, $65.03°$, and $82.36°$, which correspond to the crystallographic planes (110), (200), and (211) of the body-centered cubic

Effect of silica morphology on rheological properties 169

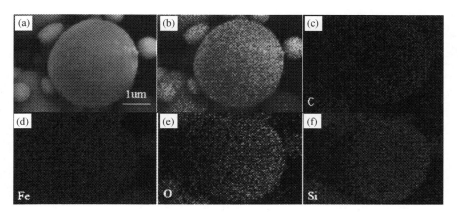

Fig. 3. FE-SEM images and EDX elemental mapping of C, Fe, O, and Si(a–f).

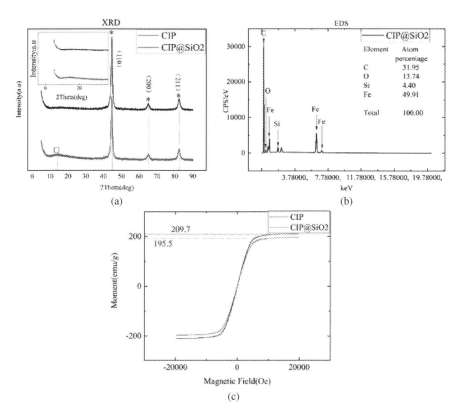

Fig. 4. CIP@SiO$_2$ magnetic particles XRD (a), EDS (b), VSM (c).

structure (JCPDS No. 87–0721), respectively.[19] Additionally, a single diffraction peak is observed for CIP@SiO$_2$ at $2\theta = 14.6°$, indicating the presence of an amorphous phase. Notably, the cladding process did not introduce any impurity peaks in the CIP@SiO$_2$ pattern, suggesting that the original structure of CIP remained unaffected by the cladding process. Figure 4(b) displays the energy dispersive spectrometer (EDS) of CIP@SiO$_2$ magnetic particles. The inherent elements Fe, C, and O in the carbonyl iron powder are present in higher quantities. At 1.740 Kev, a peak corresponding to Si can be found. Meanwhile. Fe exhibited three peaks at 0.720 Kev, 6.140 Kev and 7.31 Kev. The interaction between the electron beam and the samples caused a change in the energy of the electrons, resulting in a shift in the peak position of Fe and the appearance of multiple peaks. Figure 4(c) illustrates the hysteresis lines of the CIP and CIP@SiO$_2$ magnetizable composite particles. Both particles exhibited narrow closed curves with similar shapes, indicating that they are soft magnetic materials. The externally applied magnetic field strengths were approximately equal when both particles reached magnetic saturation. However, the saturation magnetic induction strength of CIP was 209.7 emu/g, whereas that of CIP@SiO$_2$ decreased to 195.5 emu/g due to the presence of a silica shell, resulting in a decrease of 6.7%.

4. Rheology Test

4.1. *Magnetic rheology fluid configuration*

MR fluids containing magnetizable particles with CIP and CIP@SiO$_2$ at volumes of 25%, 30%, and 35% were studied. The additives were water-based bentonite, nanoglass powder, sodium dodecyl sulfonate, and rust remover. Bentonite, glass nanopowder and sodium dodecyl sulfate can greatly improve the anti-settling properties of MR fluid, Rust inhibitor can well prevent CIP deterioration. In this study, the free SiO$_2$ particle size was 3 μm because the MR fluid with the same size of magnetic particles and abrasive particles has better mechanical properties.[20] The specific ratios are shown in Table 1.

Table 1. Composition ratio of MR fluid.

Magnetizable particles	Magnetic particle content	Additives	Deionized water
CIP	25%, 30%, 35%	Water-based bentonite 2%	71%, 66%, 61%
CIP@SiO$_2$	25%, 30%, 35%	NanoGlass Powder 1%	71%, 66%, 61%
CIP/SiO$_2$	35%	Rust inhibitor 1%	61%
		Sodium dodecyl sulfate 2%	

4.2. *Performance testing*

The evaluation of MR fluid's polishing performance was based on three main indicators: shear yield stress, viscosity of the fluid under zero magnetic field and settling stability. A lower viscosity of the fluid under zero magnetic field allows for better utilization of Newtonian hydrodynamic properties and enhances the fluidity performance. Settling stability, which indicates how slowly the fluid settles, is a crucial parameter for MR fluid. The better the settling stability, the more desirable the fluid's performance. Shear yield stress is the most significant characteristic of MR fluids because it determines their rheological performance under a magnetic field. A higher shear yield stress enables better rheological control.

The field of research on steady-state shear MR fluids tends to classify them as non-Newtonian fluids that exhibit behavior consistent with the Bingham plasticity model.[21] In the absence of an external magnetic field, MR fluid behaves like a Newtonian fluid. The relationship between shear stress and the shear strain rate of MR fluid can be described as follows:

$$\tau = \eta\gamma, \qquad (2)$$

where τ represents the shear stress of the MR fluid, η represents the viscosity coefficient of the MR fluid, and γ represents the shear strain rate of the MR fluid.

When a magnetic field is added, the MR fluid shows the characteristics of Bingham fluid. The generalized Bingham model of MR fluid is as follows:

$$\begin{cases} \tau_y = \tau_0(H)\operatorname{sgn}(\gamma) + \eta\gamma & |\tau| > |\tau_0| \\ \gamma = 0 & |\tau| \leq |\tau_0|, \end{cases} \quad (3)$$

where τ_y represents the shear stress of the MR fluid, τ_0 represents the critical shear stress of the MR fluid, H represents the magnetic field intensity of the MR fluid, γ represents the shear strain rate of the MR fluid, and η represents the viscosity coefficient of the MR fluid after reaching the yield zone.

The schematic of detection is shown in Fig. 5.

4.2.1. *Effect of linear magnetic field variation on MR fluid*

The volume fractions of magnetizable particles in the configured MR fluid were 25%, 30%, and 35%. In this experiment, the rheometer controlled the change in magnetic field intensity by controlling the current within the range of 0–4A, corresponding to a magnetic field range of 0–855 mT. Figure 6 shows that when the shear rate was held constant at 300 (1/s) and the magnetic field varied linearly from 0 mT to 855 mT, the MR fluid with different

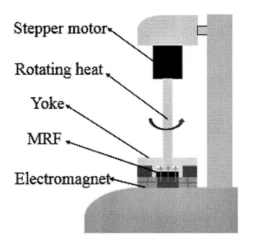

Fig. 5. Schematic of detection.

Effect of silica morphology on rheological properties

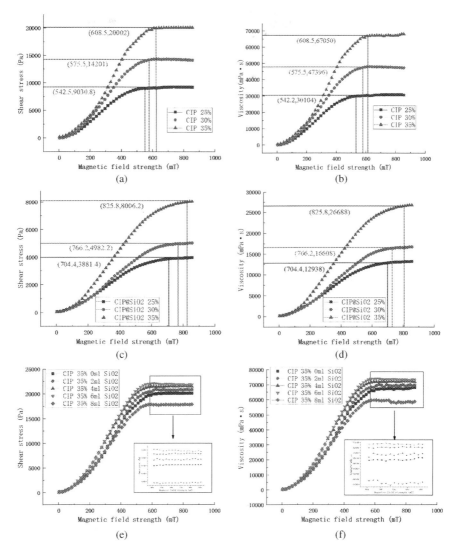

Fig. 6. Effect of linear magnetic field variation on MR fluid. (a) Variation of shear stress with magnetic in CIP MR fluid. (b) Variation of CIP MR fluid viscosity with magnetic. (c) Variation of shear stress with magnetic in CIP@SiO$_2$ MR fluid. (d) Variation of CIP@SiO$_2$ MR fluid viscosity with magnetic. (e) Variation of shear stress with magnetic in CIP/SiO$_2$ MR fluid. (f) Variation of CIP/SiO$_2$ MR fluid viscosity with magnetic.

compositions exhibited a similar trend. They all gradually increased to a certain value and then remained constant. However, significant differences were found in the maximum value reached and the magnetic field at which the maximum value was reached.

Figures 6(a) and 6(b) show the test results of CIP MR fluid. At a magnetic field of 542.5 mT, the stability values reached by the MR fluid with 25% CIP content were 9030.8 Pa and 30104 mPa $\cdot$ s. For 30% CIP content, the values were 14,201 Pa and 47,396 mPa $\cdot$ s at a magnetic field of 575.5 mT. The largest values were observed with 35% CIP content, reaching 20,002 Pa and 67,0504 mPa $\cdot$ s at a magnetic field of 608.5 mT. These results indicated that the magnetic field required to reach magnetic saturation of the MR fluid became stronger as the CIP particle content increased, and the maximum shear stress and viscosity that can be achieved increased significantly.

Figures 6(c) and 6(d) present the test results of CIP@SiO$_2$ MR fluid. While the trend of alteration remains unaltered compared to CIP MR fluid, the overall performance of the MR fluid reduced significantly. As the magnetic field increased towards a stable state, the attainable stable values decreased. For the MR fluid with 25% CIP@SiO$_2$ content, the shear stress and viscosity reached their maximum at a magnetic field of 704.4 mT, with values of 3881.4 Pa and 12938 mPa $\cdot$ s, respectively. Similarly, for the MR fluid with 30% CIP@SiO$_2$ content, the shear stress and viscosity reached their maximum at a magnetic field of 766.2 mT, with values of 4982.2 Pa and 16,608 mPa $\cdot$ s, respectively. Finally, for the MR fluid with 35% CIP@SiO$_2$ content, the shear stress and viscosity reached their maximum at a magnetic field of 825.8 mT, with values of 8006.2 Pa and 26,688 mPa $\cdot$ s, respectively.

Figures 6(e) and 6(f) show the test results for CIP/SiO$_2$ MR fluid. The initial shear stress and viscosity of each MR fluid were approximately equal with the gradual increase in SiO$_2$ abrasive content from 0 vt%, 2 vt%, 4 vt%, and 6 vt% to 8 vt%. However, as the magnetic field increased, the difference became more apparent. The magnetic field required to reach magnetic saturation was

Effect of silica morphology on rheological properties 175

approximately 608.5 mT, and the shear stress and viscosity reached their maximum values when the SiO_2 content was 6% at 22,103 Pa and 73,977 mPa·s, respectively. However, the shear stress and viscosity decreased significantly after the particle content increased to 8 vt%. When the magnetic particle content was the same, the addition of SiO_2 can improve the mechanical properties of the MR fluid, and the maximum shear stress and viscosity increased However, no significant change was found in the maximum magnetization magnetic field, indicating that the magnetizable particles are the key factors in the rheological properties of the MR fluid.

4.2.2. *Effect of linear shear rate change on MR fluid*

When the magnetic field was kept constant at 465 mT (current kept at 2 A) and the shear rate was increased linearly from 0 r/min to 800 r/min, the shear stress and viscosity trends of different MR fluids were similar. The shear stress increased with the increase in shear rate, whereas the viscosity decreased. The change trend was rapid at first and then gradual.

The CIP content is 25%, 30%, and 35%, as shown in Figs. 7(a) and 7(b). When the shear rates were equal, the shear stress of the CIP MR fluid increased with the increase in magnetizable particle content. At a shear rate of 0, the initial shear stress increased with the increase in CIP content, measuring 7550.6 Pa, 8270.7 Pa, and 11,810 Pa. As the shear rate increased to 800 (1/s), the shear stress of the three types of CIP MR fluid increased to 1567.2 Pa, 2341.3 Pa, and 2438 Pa. The final values were positively correlated with the CIP content, measuring 9117.8 Pa, 10,612 Pa, and 14,248 Pa.

The concentration of magnetizable particles in $CIP@SiO_2$ MR fluid varied at 25%, 30%, and 35%, as shown in Figs. 7(c) and 7(d). When the shear rate increased linearly, the shear stress gradually increased, whereas the viscosity gradually decreased. This trend exhibited a rapid change followed by a slower change. Under the same shear rate, the shear stress and viscosity increased with an increase in the content of $CIP@SiO_2$ particles. Comparatively,

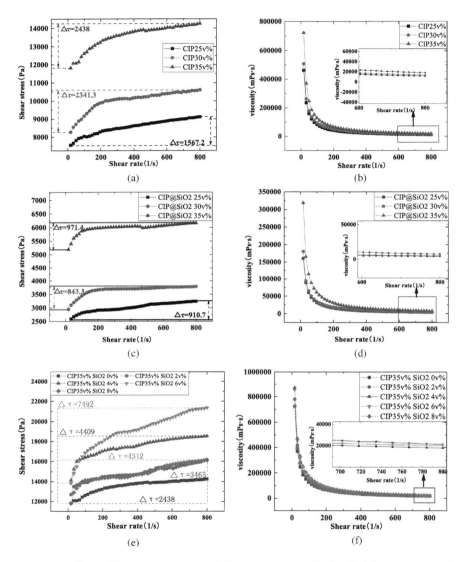

Fig. 7. Effect of linear variation of shear rate on MR fluid. (a) Variation of shear stress with Shear rate in CIP MR fluid. (b) Variation of CIP MR fluid viscosity with shear rate. (c) Variation of shear stress with Shear rate in CIP@SiO$_2$ MR fluid. (d) Variation of CIP@SiO$_2$ MR fluid viscosity with shear rate. (e) Variation of shear stress with Shear rate in CIP/SiO$_2$ MR fluid. (f) Variation of CIP/SiO$_2$ MR fluid viscosity with Shear rate.

Effect of silica morphology on rheological properties 177

under the same conditions, the mechanical properties of CIP@SiO$_2$ MR fluid were found to be weaker than those of CIP MR fluid.

The concentration of CIP particles remained constant at 35% and different amounts of free SiO$_2$ abrasive particles were added. Figures 7(e) and 7(f) evidently show that the shear stress increased with the increase in shear rate, whereas the viscosity decreased. Under the same conditions, the shear rate and viscosity reached their maximum at a SiO$_2$ content of 6%. However, when the SiO$_2$ content was further increased to 8%, a significant decrease was found in shear rate and viscosity. This finding indicated that increasing the content of free SiO$_2$ abrasive particles can enhance the mechanical properties of the MR fluid within a certain range when a constant magnetic field and a linear increase in shear rate are applied. However, when the content of free SiO$_2$ exceeds a certain threshold, it has an opposite effect on the mechanical properties of MR fluid.

4.2.3. *Effect of linear change of shear rate on properties of MR fluid under zero magnetic field*

Figure 8 shows the test curves depicting the mechanical properties of the MR fluid under zero magnetic field conditions. The shear rate ranged from 0 to 300 (1/s), and it increased linearly. The results indicated that the shear stress increased as the shear rate increased, whereas the viscosity decreased. Notably, compared with those under magnetic field condition, the shear stress and viscosity demonstrated a significant decrease in the absence of a magnetic field.

As observed in Figs. 8(a) and 8(b), when no magnetic field was present, the shear stress of the CIP MR fluid increased linearly with the shear rate, whereas the viscosity decreased. Additionally, under the same shear rate condition, the shear stress and viscosity increased with an increase in iron powder content. When the shear rate ranged from 0 to 300 (1/s), and the change in shear stress increased with the increase in iron powder content. Specifically, for

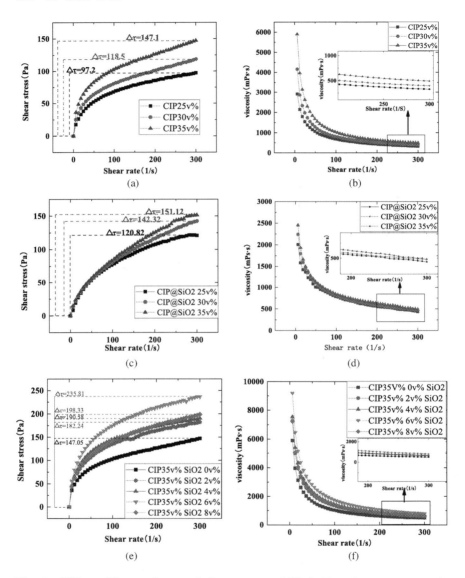

Fig. 8. Effect of linear change of shear rate on MR fluid under zero magnetic field. (a) Variation of shear stress with shear rate in CIP MR fluid. (b) Variation of CIP MR fluid viscosity with shear rate. (c) Variation of shear stress with shear rate in CIP@SiO$_2$ MR fluid. (d) Variation of CIP@SiO$_2$ MR fluid viscosity with shear rate. (e) Variation of shear stress with shear rate in CIP/SiO$_2$ MR fluid. (f) Variation of CIP/SiO$_2$ MR fluid viscosity with shear rate.

CIP contents of 25%, 30%, and 35%, the shear stress values were 97.2 Pa, 118.5 Pa, and 147.1 Pa, respectively.

As shown in Figs. 8(c) and 8(d), the shear stress of $CIP@SiO_2$ MR fluid increased as the shear rate was increased, whereas the viscosity decreased under zero magnetic field conditions. Additionally, under the same conditions, the shear stress and viscosity of $CIP@SiO_2$ MR fluid were higher than those under the same concentration of CIP MR fluid.

As shown in Figs. 8(e) and 8(f), when the CIP content was 35%, adding different amounts of SiO_2 abrasive particles, when the SiO_2 particles increase to 6%, the shear stress and viscosity reached the maximum. When the amount of SiO_2 abrasive particles was increased to 8%, the shear stress and viscosity showed a substantial reduction. Therefore, under the condition of zero magnetic field, the shear rate increased linearly, in a certain range to increase the content of SiO_2 abrasive particles demonstrated a greater improvement on the mechanical properties of the MR fluid, with the SiO_2 content increasing again, the performance of the MR fluid decreased.

The above experiments demonstrated that magnetizable particles are the decisive factor affecting the rheological properties of MR fluid, and the higher the content of magnetizable particles, the stronger the magnetic field required to reach magnetic saturation, and the greater the shear stress and viscosity. The presence of SiO_2 in the form of shells can improve the mechanical properties of the MR fluid at zero magnetic field. However, after a magnetic field was applied, the response of $CIP@SiO_2$ magnetizable particles to the magnetic field became sluggish. Moreover, the magnetic field needed to reach magnetic saturation increased and the shear stress and viscosity decreased due to the hindering effect of SiO_2 shells on the magnetic field. The free SiO_2 abrasive particles can greatly improve the performance of the MR solution and they did not change the magnetic field size when the MR solution reached magnetic saturation. Furthermore, the SiO_2 content was optimal at 6%. In summary, the different states of silica present have different effects on the performance of the MR solution.

5. Stability Experiments and Conclusions

The MR fluid behaves as a low-viscosity Newtonian fluid when no magnetic field is present. In this scenario, the magnetizable particles are assumed to be spherical and the interaction between the particles during the settling process is disregarded. The main forces acting on the magnetizable particles are gravity and buoyancy because they have different densities. Consequently, the magnetizable particles experience significant settling relative to the carrier fluid.[22]

$$G = \frac{4}{3}\pi\left(\frac{d}{2}\right)^3 g\rho = \frac{\pi}{6}\rho g d^3, \tag{4}$$

$$F = \frac{4}{3}\pi\left(\frac{d}{2}\right)^3 g\rho_0 = \frac{\pi}{6}\rho_0 g d^3, \tag{5}$$

$$f = A\varepsilon\frac{\rho_0 u^2}{2}, \tag{6}$$

G represents the gravitational force on the magnetizable particle, F represents the buoyancy force on the magnetizable particle, f represents the motion resistance, d represents the average particle size of magnetizable particles, ρ represents the density of magnetizable particles, ρ_0 represents the density of liquid carrier, g represents the gravitational acceleration, A represents the surface area of magnetizable particles, and u represents the sedimentation rate of magnetizable particles.

It can be seen from Tables 2, 3 and 4, the supernatants of CIP@SiO_2, CIP, and CIP/SiO_2 after sedimentation, were pure white, yellow, and light yellow, respectively. This observation suggested that during the long-term sedimentation process, the CIP in the CIP MR fluid underwent a reaction with water and air, leading to increased rust formation. Meanwhile the CIP@SiO_2 MR fluid, remained pure white due to the uniform coating of SiO_2 on the surface of CIP, effectively preventing its oxidation reaction. Additionally, the inclusion of free SiO_2 abrasive particles in the CIP/SiO_2 MR fluid helped reduce the oxidation reaction of CIP, resulting in light-yellow supernatant.

The results from Figs. 9 and 10 shown that the stability of the MR fluid improved with an increase in magnetic particle content.

Table 2. Sedimentation change of CIP@SiO$_2$ MR fluid.

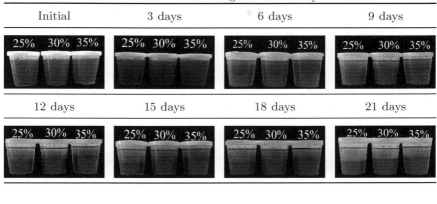

Table 3. Sedimentation change of CIP MR fluid.

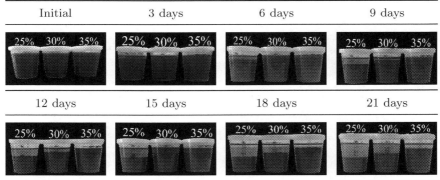

Table 4. Sedimentation change of CIP/SiO2 MR fluid.

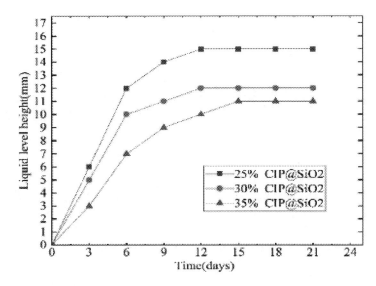

Fig. 9. CIP@SiO$_2$ MR fluid sedimentation change curve.

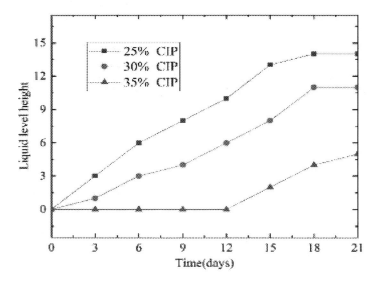

Fig. 10. CIP MR fluid sedimentation change curve.

When the magnetizable particle content in the CIP@SiO$_2$ MR fluid reached 35%, the fluid surface height remained unchanged after 15 days, indicating the completion of the settling motion. The settling motion took significantly longer time than in the MR fluid with 25% and 30% magnetizable particles and the anti-settling properties of the CIP MR fluid significantly improved when the magnetizable particle content was the same compared with those of the CIP@SiO$_2$ MR fluid. When free abrasive particles were added, the anti-settling properties further improved.

6. Conclusion

In accordance with the above experimental results, the following conclusions can be drawn:

The sol–gel method was adopted to form a uniform shell on the surface of CIP by using TEOS as the silicon source.

The rheological effect of MR fluid improves with an increase in magnetizable particle content. The state of SiO$_2$ does not affect the rheological effect of the MR fluid, but different states of SiO$_2$ have a significant effect on the rheological performance. Under zero magnetic field conditions, the SiO$_2$ shell improves the mechanical properties of the MR fluid. Under magnetic field conditions, the rheological effect of CIP@SiO$_2$ MR fluid is significantly weaker than that of CIP MR fluid. The free SiO$_2$ abrasive particles have a noticeable improvement effect on the rheological effect of the MR fluid, reaching the optimum at a volume fraction of 6%.

The stability of MR fluid improves as the magnetizable particle content increases. Under the same concentration of magnetizable particles, the stability of CIP MR fluid is superior to that of CIP@SiO$_2$ MR fluid. The state of SiO$_2$ has a significant effect on the stability of the MR fluid. When it exists in the form of a shell, it weakens the stability of the MR fluid, whereas it greatly improves the stability of the MR fluid when it exists in a free state. No apparent sedimentation phenomenon occurs within the same time.

Acknowledgments

This project has been supported by: the Natural Science Foundation of Fujian Province (grant no.2020J01874,2023J01342), the Program for Innovative Research Team in Science and Technology in Fujian Province University (2020No.Grant:No.12). Fujian Provincial Key Project of Science and Technology Innovation (2022G02007) and High-level talents foundation of Fuzhou Polytechnic. National Natural Science Foundation (52275413).

ORCID

Bingsan Chen ◉ https://orcid.org/0000-0001-6723-9305

References

1. M. Cvek *et al.*, *J. Rheol.* **60**, 687 (2016).
2. C. S. Maurya *et al.*, *Rheol Acta* **61**, 99 (2022).
3. Gordaninejad *et al.*, *J. Struct. Eng.* **136**, 135 (2010).
4. G. Liu *et al.*, *Smart Mater. Struct.* **31**, 043002 (2022).
5. H. Eshgarf *et al.*, *J Energy Stor.* **50**, 104648 (2022).
6. P. K. Agyeman *et al.*, *Case Stud.* **28**, 101686 (2021).
7. W. Yu–yue *et al.*, *Appl. Surf. Sci.* **360**, 224 (2016).
8. C. S. Maurya *et al.*, *Mater. Sci. Forum* **1060**, 141 (2022).
9. A. J. F. Bombard *et al.*, *Smart Mater. Struct.* **23**, 125013 (2014).
10. R. Milde *et al.*, *Int. J. Mol. Sci.* **23**, 12187 (2022).
11. A. Kaide *et al.*, *J. Chem. Eng. Jpn* **53**, 438 (2020).
12. B. Chen *et al.*, *Funct. Mater. Lett.* **296**, 1145 (2021).
13. J. Xu *et al.*, *Colloid Polym. Sci.* **296**, 1145 (2018).
14. A. J. F. Bombard *et al.*, *Int. J. Mod. Phys. A* **21**, 0704576 (2007).
15. M. N. Aruna *et al.*, *J. Magn. Magn. Mater.* **529**, 167910 (2021).
16. D. Shao *et al.*, *J. Colloid Interface Sci.* **336**, 526 (2009).
17. B. Jia *et al.*, *Scr. Mater.* **56**, 677 (2007).
18. B. Luo *et al.*, **26**, 1674 (2021).
19. S. Li *et al.*, *Trans. Nonferr. Metal Soc.* **30**, 3067 (2020).
20. L. Nagdeve *et al.*, *Mach. Sci. Technol.* **22**, 493 (2018).
21. M. Cvek *et al.*, *J. Rheol.* **60**, 687 (2016).
22. N. Wang *et al.*, *J. Magn. Magn. Mater.* **501**, 166443 (2020).

Chapter 11

Porous pyroelectric material for waste heat harvesting

Qingping Wang[*,‡,¶], Haifeng Luo[*], Zhanxiang Xu[†] and Tian Wu[*,§,¶]

Department of Physics & Mechanical and Electronic Engineering
Hubei University of Education Wuhan 430205, P. R. China
†*School of International Education (Engineering Management)*
Tianjin Chengjian University, Tianjin 300072, P. R. China
‡*qw608@bath.ac.uk*
§*twu@whu.edu.cn*

The utilization of pyroelectric materials to harvest waste heat from the ambient environment has gained much interest. However, pyroelectric energy harvesting technology still faces the challenge of low output voltage. In this paper, a porous pyroelectric ceramic with 0.3 wt.% poly(methyl methacrylate) (PMMA) as the additive was prepared via a conventional solid-state reaction technique. Our experimental results indicated that the porous PMN–PMS–PZT: 0.3PMMA ceramic had a decrease in both relative permittivity and volume heat capacity at a temperature of 44°C, compared to that of the dense sample. Additionally, the porous ceramic demonstrated an enhanced voltage responsivity from 0.50 to 0.52 m^2/C and a higher figure of merit from 0.64 to 0.65 nm^3/J for energy harvesting, compared to the dense sample. When 0.3 wt.% PMMA was added, the voltage generated by the device increased by 17.01% due to the presence of pores, which suggests

---

¶Corresponding author.

To cite this article, please refer to its earlier version published in the Functional Materials Letters, Volume 16(7), 2340036 (2023), DOI: 10.1142/S1793604723400362.

186   *Q. Wang et al.*

a potential application in the field of waste heat harvesting and sensing.

Keywords: Pyroelectric material; porosity; figure of merits; waste heat harvesting.

1. Introduction

To address the energy crisis and global warming, researchers have come up with a variety of measures to curb these problems.[1,2] Among these solutions, developing innovative materials and structures that can effectively harness waste heat has gained significant interest.[3-12] Pyroelectric energy harvesting technique is able to convert heat to electrical energy, providing a clean and environmentally friendly source of power.[13-16] This approach utilizes the pyroelectric effect, which involves tailoring the level of polarization of pyroelectric materials by subjecting them to cyclic heating and cooling. This process generates the electrical charge and it is possible to accumulate the charge in a storage capacitor during thermal cycling, or to allow the charge to flow through an external electrical circuit to harvest power.[9,10,17,18] The short circuit electric current of pyroelectric materials under the thermal cycling can be expressed as follows:

$$I = A \cdot P \cdot (dT/dt), \tag{1}$$

where A is the effective electrode area, p is the pyroelectric coefficient and dT/dt is the rate of change in temperature.[19] Equation (1) implies that, if A is given, to improve the generated current, the pyroelectric material should have larger p and dT/dt. The open circuit voltage generated from the pyroelectric materials can be written as follows[20]:

$$V = \frac{p}{\varepsilon_0 \varepsilon_r} \cdot \Delta T \cdot t, \tag{2}$$

where ε_0 and ε_r are the permittivity of vacuum and relative permittivity, ΔT is the temperature change and t is the thickness of the sample. Equation (2) shows that, to maximize the output voltage, the pyroelectric material should have larger pyroelectric coefficient and smaller relative permittivity. Meanwhile, the output current and voltage and energy harvesting figure of merit are often determined by these equations[20,21]:

$$\text{Current responsivity,} \quad F_I = \frac{p}{C_E},\tag{3}$$

$$\text{Voltage responsivity,} \quad F_V = \frac{p}{\varepsilon_r \varepsilon_0 C_E}.\tag{4}$$

$$\text{Energy harvesting figure of merit:} \quad F_E = \frac{p^2}{\varepsilon_r \varepsilon_0 (C_E)^2},\tag{5}$$

where C_E is the volume heat capacity.

These equations show that, to enhance these indicators, we should increase ρ and decrease ε_r and C_E. Generally, increasing ρ can be achieved by microstructure engineering.[18,19,22] structure design of the devices[23,24] and external bias electrical field.[25,26] Reducing ε_r and C_E can be accomplished by the introduction of pores in the structure of the materials.[17,27] Previous reports show that pyroelectric materials with porous structures have higher figure of merits compared to the dense material as porosity can decrease the ε_r and C_E.[17,28 30] However, the introduction of porosity also reduces ρ simultaneously and the advantages in performance can only be achieved if p is not significantly decreased. This intricate link between thermal and dielectric properties means that tuning the porosity allows the energy harvester to achieve an optimal pyroelectric response.

To date, a variety of approaches have been used to prepare porous pyroelectric materials such as partial sintering, direct foaming, replica template and sacrificial template.[31] Sacrificial template is a highly popular method in the production of porous materials. This is due to its versatility, as there is a wide range of different shapes and sizes of pore-formers available, which can be added to the ceramic compact and then removed by heat treatment, creating the corresponding pores. This method used to prepare the porous ceramics resulted in a range of porosity from 10% to 90% and an average pore size ranging from 1 to 700 μm.[31] Poly(methyl methacrylate) (PMMA) was therefore selected as a forming agent to reduce the p and C_E and improve the figure of merits since this technique offers an easy and cheap way of creating porosity.[28]

Experimental results show that ε_r and C_E were reduced and the voltage responsivity and F_E as well as the output performance of the pyroelectric energy harvester were improved. This work therefore showcases the potential application in the field of energy harvesting and sensing.

2. Experimental Procedure

2.1. *Preparation of pyroelectric ceramic*

Pyroelectric ceramics of $Pb[(Mn_{1/3}Nb_{2/3})_{1/2}(Mn_{1/3}Sb_{2/3})_{1/2}]_{0.04}(Zr_{0.95}Ti_{0.05})_{0.96}O$ (PMN–PMS–PZT) and PMN–PMS–PZT: 0.3PMMA were produced using the conventional solid-state reaction as reported.[17] The raw materials used for the dense ceramic remain unchanged,[17] while the raw materials for the porous ceramic include an addition of 0.3 wt.% PMMA to the dense raw material. Before the ball milling, a 10 mol.% excess of PbO was added to make up for the loss of lead and to ensure that the ceramics didn't form the pyrochlore phase. After 4 h of ball milling, the mixture was subjected to a temperature of 700°C for 3 h. After completing the above calcination process, the polycrystalline powder was pressed into discs with a diameter of 10 mm. This was done using a pressure of 8 MPa and the binder used was polyvinyl acetate (PVA). The discs were subsequently subjected to the sintering process at a temperature of 1230°C for a duration of 2 h in the presence of air. The disc specimens were polished with the thickness of 0.35 mm and silver electrodes were applied on both surfaces of the specimens using screen-printing method. Afterwards, the ceramics underwent polarization by applying a direct current (DC) field of 3.0 kV/mm, while immersed in a silicone oil bath at a temperature of 90°C. Subsequently, the ceramics were left to age for a period of 24 h.

2.2. *Characterization*

The microstructure of the samples was analyzed using field emission scanning electron microscopy (FE-SEM Sirion 200, FEI). The piezoelectric response of the porous sample was obtained using an atomic force microscope with a high voltage module, the process of preparing the sample and using the necessary equipment was identical to what was previously described.[17] The dielectric property was examined by an impedance analyzer (Agilent4294A, Agilent Technology Inc., CA). The ferroelectric hysteresis loops were obtained by a ferroelectric system (CPE1601, PolyK Technologies, State College, PA, USA). The pyroelectric property was analyzed using the standard Byer–Roundy method. The volume heat capacity (C_E) was calculated using the following equation:

$$C_E = \rho C_{p(\text{dense})}(1 - \varphi), \tag{6}$$

where $C_{p(\text{dense})} = 0.29$ J/gK.[17] ρ and φ are the density of the samples and porosity faction, respectively. The density measurement of the samples was obtained by utilizing the Archimedes' method.

For energy harvesting measurement, the temperature variation and rate of the change in temperature were obtained by a thermal controller and two Peltier modules, respectively. The output performance of the pyroelectric energy harvesters was measured via a data acquisition system that relied on LabVIEW software.

3. Results and Discussion

The image in Fig. 1(a) demonstrates that the dense ceramic material possesses a surface morphology that is both fine-grained and uniform. No microcracks or other defects in the microstructure were observed in the ceramics. Figure 1(b) shows that there are pores and defects appearing on the surface of the porous PMN–PMS–PZT: 0.3PMMA ceramic, the inset presents the surface morphology of the forming agent PMMA. The corresponding cross section morphology of the porous ceramic is illustrated in Fig. 1(c). Different sizes of pores were shown in the ceramic matrix. To gain insights into the switching behavior within the local domain, a thorough examination of the amplitude and phase of the porous material has been conducted. The results of this analysis are illustrated in Figs. 1(d) and 1(e). Specifically, when no bias voltage is applied to the porous ceramic, the domains with varying angles are

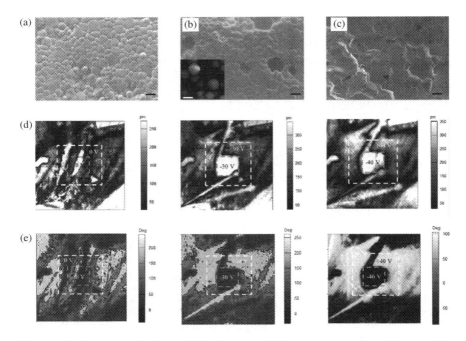

Fig. 1. Surface morphology of (a) dense and (b) porous ceramics. The inset is the SEM of PMMA, scale bar: 2 μm. (c) cross section SEM of porous ceramic, scale bar: 2 μm. (d) piezoelectric force microscopy amplitude and (e) phase images of porous ceramics under various applied voltages such as 0 V, ±30 V and ±40 V.

randomly distributed. However, when a voltage of −30 V is applied, the domains undergo a noticeable switching effect. This effect is characterized by larger amplitude and phase compared to the original state. In phase images as shown in Fig. 1(e), there is a clear distinction in the orientation of the domains, with certain areas exhibiting a 180° contrast and others showing a non-180° contrast. Increasing the applied voltage further to −40 V results in a stronger response in both amplitude and phase, indicating that more domains have reversed their orientation.

Figures 2(a)–2(d) present the variations of dielectric properties of the dense and porous ceramics as function of the temperature from room temperature (RT) to 300°C at different working frequencies. Figures 2(a) and 2(c) illustrate that each ceramic material exhibits a single dielectric peak.

Porous pyroelectric material for waste heat harvesting 191

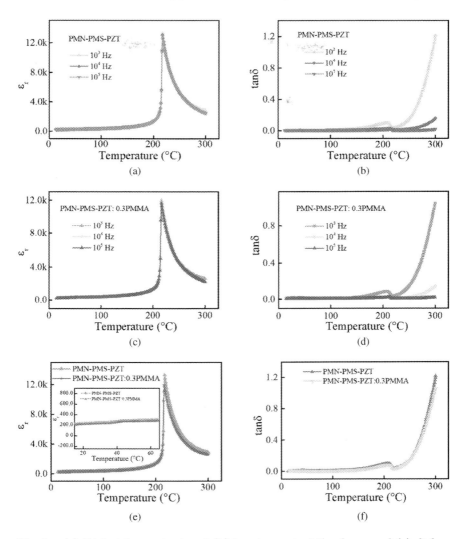

Fig. 2. (a) Dielectric constant and (b) loss tangent of the dense and (c) dielectric constant and (d) loss tangent of the porous sample with temperature from room temperature to 300°C at different frequencies. (e) comparison of dielectric constants and (f) loss tangent with temperature from room temperature to 300°C at 1 kHz.

Additionally, the dielectric constant of the ceramics is observed to rise as the temperature increases, reaching its highest value at the Curie point. Subsequently, the dielectric constant sharply decreases with a further increase in temperature, indicating the ceramics undergo a phase transition. Specifically, below the Curie point, the ceramics exhibit a ferroelectric phase, while above the Curie point, they lose this property and become paraelectric. Additionally, the dielectric constant of both the dense and porous ceramics decreases as the operating frequency increases, which can be attributed to the varying response time of the dipoles.[32] The loss tangent of both ceramics shows the similar changing trend with increasing frequency and was less than 0.02 at 70°C. To calculate the figure of merits, we also compared the dielectric constants of the dense and porous ceramics at 44°C at 1 kHz as shown in Fig. 2(e) and they are 295 and 272 for the dense and porous samples, respectively, see Table 1. The loss tangent of the porous ceramic is smaller than that of the dense sample at temperature below 70°C at 1 kHz.

Figure 3(a) depicts the hysteresis loops of the dense and porous ceramics at RT, showcasing the relationship between polarization and electric field. When the content of PMMA was increased from 0% to 0.3 wt.%, both exhibited typical ferroelectric properties, with the saturation polarization decreasing from 22.35 to 21.53 $\mu C/cm^2$, which can be attributed to the presence of pores that generates a depolarization field.[17] Similarly, the residual exhibits the same trend, see Table 1, which is mainly due to the depolarization resulted in the electric field around the pores.[17] In addition, the coercive field also decreased with an addition of 0.3 wt.% PMMA,

Table 1. Parameters of the dense and porous pyroelectric ceramics.

Parameters Materials	P_s	P_r	E_c	p ($\mu C/$ m^2k) at 44°C	ε_r at 44°C	C_E (J/cm^{3}K)	F_i $(10^{-10}m/V)$ at 44°C	F_v (m^2/C) at 44°C	F_E (nm^3/J) at 44°C
PMN-PMS-PZT	22.35	6.49	−15.85	29.69	295	2.29	12.94	0.50	0.64
PMN-PMS-PZT:0.3PMMA	21.53	5.64	−15.01	27.63	272	2.20	12.55	0.52	0.65

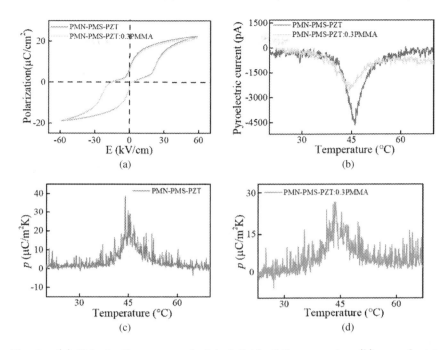

Fig. 3. (a) Polarization versus electrical field of the ceramics. (b) pyroelectric current. (c) and (d) pyroelectric coefficient versus temperature for the dense and porous ceramics over the temperature range of 25°C–65°C.

which can be explained that the pores reduce the stiffness of the ceramic matrix and in turn promoting domain switching. The pyroelectric current of the dense and porous ceramics over the temperature of 25–65°C is presented in Fig. 3(b), the corresponding pyroelectric coefficient calculated according to Eq. (1) can be seen in Figs. 3(c) and 3(d), respectively. The detailed value decreased from 29.69 $\mu C/m^2 K$ to 27.63 $\mu C/m^2 K$ at 44°C, see Table 1, which is similar to the changing trend in saturation polarization. Based on these values and Eqs. (3)–(5), the figure of merits can be obtained as shown in Table 1. At 44°C, the current responsivity decreased while the voltage responsivity and energy harvesting figure of merit both increased, indicating that the porous ceramic sample has a stronger pyroelectric response at the same thermal field compared to the dense sample.

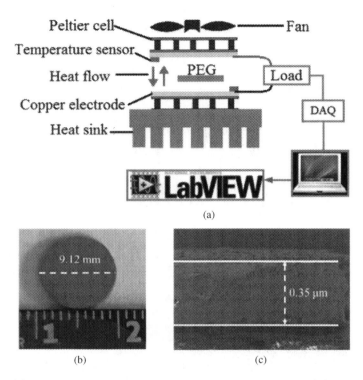

Fig. 4. (a) Schematic of pyroelectric energy harvesting system, (b) photographs of the diameter and (c) thickness of the ceramics.

To investigate the pyroelectric energy harvesting properties of the ceramics, a test system was set up as seen in Fig. 4(a). Two Peltier modules were used to provide heating and cooling for the samples. Two K-type thermocouples were utilized to observe and track the temperature of both the hot and cold flows of the samples. A heat sink and a fan were selected to effectively cool the pyroelectric energy generators. The performance results of the devices can be obtained via the LabVIEW software. The samples had a diameter of 9.12 mm and a thickness of 0.35 μm, as depicted in Figs. 4(b) and 4(c). The temperature variation between 25°C and 50°C, as well as the rate of change in temperature in the energy harvesters using the dense and porous ceramics, is illustrated in

Porous pyroelectric material for waste heat harvesting 195

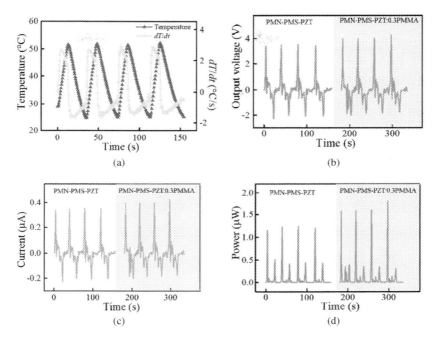

Fig. 5. (a) Cycling temperature and rate of change in temperature of the ceramics. (b) Voltage generated, (c) current across the 10 MΩ resistor and (d) power harvested.

Fig. 5(a). The generated voltage, current and power across the resistive load of 10 MΩ can be observed in Figs. 5(b)–5(d), respectively. By analyzing Fig. 5(b), it is evident that the voltage produced by the energy harvester using the dense ceramic is approximately 3.396 V, while the one using the porous ceramic generated around 3.975 V. This indicates an increase of 17.1% in voltage, suggesting a more enhanced pyroelectric response. At a load resistance of 10 MΩ, the current values of the two devices are 0.340 and 0.398 μA, respectively. The corresponding power values of the devices are 1.152 and 1.582 μW, respectively. The device using the porous ceramic showed a 37.33% improvement over the device using the dense sample, which indicates that the addition of PMMA is effective in improving the heat-to-electricity conversion.

4. Conclusion

We carefully designed and prepared the porous pyroelectric ceramic to achieve a higher conversion from heat to electricity. The performance of the porous ceramic is enhanced in comparison to that of the dense sample. This improvement can be attributed to the presence of pores in the ceramic matrix, which leads to a reduction in the dielectric constant and volume heat capacity. As a result, the porous ceramic demonstrates higher voltage responsivity and energy harvesting figure of merit. The pyroelectric energy harvester with porous PMN–PMS–PZT: 0.3 PMMA ceramic was able to generate 3.975 V under the thermal cyclic condition, increasing by 17.1% compared to the device with the dense sample. This project showcases a possible utilization in the domain of converting thermal energy into electrical energy.

Acknowledgments

Wang acknowledges the support of National Nature Science Foundation of China (Grant No. 51902094) and the Postdoctoral Scholarship from Chinese Scholarship Council (Grant No. 201908420009).

References

1. F. Belaïd *et al.*, *Renew. Energy* **205**, 534 (2023).
2. M. Farghali *et al.*, *Environ. Chem. Lett.* **21**, 2003 (2023).
3. A. Rafique *et al.*, *Nano-Micro Lett.* **15**, 40 (2023).
4. J. Margielewicz *et al.*, *Energy Convers. Manag.* **277**, 116672 (2023).
5. H. Yu *et al.*, *Energy Convers. Manag.* **287**, 117106 (2023).
6. K. Maity *et al.*, *ACS Appl. Mater. Interfaces* **15**, 13956 (2023).
7. C. Xu *et al.*, *Energy Environ. Sci.* **16**, 983 (2023).
8. H. Wang *et al.*, *Nano Energy* **108**, 108184 (2023).
9. H. S. Choi *et al.*, *Appl. Energy* **344**, 121271 (2023).
10. R. Mondal *et al.*, *Mater. Today* **66**, 273 (2023).
11. S. Hur *et al.*, *Nano Energy* **114**, 108596 (2023).

12. M. Gao *et al.*, Optimization of core size and harvested power for magnetic energy harvesters based on cascaded magnetics, *IEEE Appl. Power Electron. Conf. Expo.*, Orlando, Florida, USA, 2023, pp. 2926–2932.
13. Z. Zheng *et al.*, *Nat. Sustain.* **6**, 180 (2022).
14. S. Pandya *et al.*, *NPG Asia Mater.* **11**, 26 (2019).
15. A. Thakre *et al.*, *Sens.* **19**, 21170 (2019).
16. A. Kumar *et al.*, *J. Korean Ceram. Soc.* **56**, 412 (2019).
17. Q. Wang *et al.*, *Nano Energy* **102**, 107703 (2022).
18. H. Li *et al.*, *Nano Energy* **102**, 107657 (2022).
19. Q. Wang *et al.*, *J. Alloys Compd.* **710**, 869 (2017).
20. C. R. Bowen *et al.*, *Energy Environ. Sci.* **7**, 3836 (2014).
21. C. R. Bowen *et al.*, *Mater. Lett.* **138**, 243 (2015).
22. W. Dong *et al.*, *Adv. Sci.* **9**, e2105368 (2022).
23. J. Lee *et al.*, *Nano Energy* **97**, 107178 (2022).
24. Y. Yang, *Nat. Energy* **7**, 1007 (2022).
25. B. A. K. Sodige *et al.*, *Energies* **16**, 4728 (2023).
26. P. Lheritier *et al.*, *Nature* **609**, 718 (2022).
27. M. Li *et al.*, *Adv. Powder Mater.* **1**, 1000032 (2022).
28. M. Aggarwal *et al.*, *Mater. Today Commun.* **31**, 103302 (2022).
29. C. Yu *et al.*, *J. Polym. Res.* **29**, 79 (2022).
30. L. Delimova *et al.*, *Mater (Basel)* **16**, 5171 (2023).
31. E. Mercadelli *et al.*, *IEEE Trans. Ultrason. Ferroelectr. Freq. Control* **68**, 217 (2021).
32. Q. Wang *et al.*, *J. Mater. Chem. A* **6**, 5040 (2018).

Chapter 12

Preparation of Mo^{6+}, Gd^{3+}-doped TiO$_2$ nanotube arrays and study of their organization and photocatalytic properties

Chaoqian Qin, Jinghong Du ◉*, Jiaxing Chen, Jiarui Yang
and Shengyang He

School of Materials Science and Engineering
Kunming University of Science and Technology
Kunming 650032, P. R. China
*cldjh@sina.com

Titanium dioxide nanotube arrays doped with Mo^{6+}, Gd^{3+}, and Mo^{6+}–Gd^{3+} were prepared by anodic oxidation combined with a two-step electrochemical method. The crystalline structure, surface morphology, light absorption properties, and crystalline transformation process were characterized using XRD, SEM, UV–Vis, DSC, and FT-IR. The results show that Mo^{6+} and Gd^{3+} single doping of TiO$_2$ can promote the transformation of amorphous TiO$_2$ to anatase and inhibit the transformation of anatase phase to rutile, but Mo^{6+}–Gd^{3+} co-doping promotes both phase transformations, and the anatase compositions of the TiO$_2$ heat-treated at 450°C were all over 91%; The tube diameters of TiO$_2$ nanotubes were all around 60–70 nm; the Mo^{6+} - and Gd^{3+} -doped TiO$_2$ light absorption peaks were all significantly red-shifted, with the most obvious change in the Mo^{6+}–Gd^{3+} co-doping, where the forbidden bandwidth was reduced to 3.01 eV; The degradation rate of pure TiO$_2$

---

*Corresponding author.

To cite this article, please refer to its earlier version published in the Functional Materials Letters, Volume 16(7), 2340037 (2023), DOI: 10.1142/S1793604723400374.

photocatalytic degradation of methylene blue for 8 h was only 41.9%, while the degradation rate of Mo^{6+}–Gd^{3+} co-doped TiO_2 was as high as 91.51%; Mo^{6+}, Gd^{3+}, single doping and co-doping could increase the amount of •OH, •OOH adsorbed on the surface of TiO_2, which led to the improvement of catalytic performance.

Keywords: Titanium dioxide nanotubes; Mo^{6+} doping; Gd^{3+} doping; forbidden bandwidth; photocatalytic degradation.

1. Introduction

In the 21st century, rapid social and industrial development resulted in a high degree of water pollution, and finding an efficient and environmentally friendly material to deal with water pollution has become a top priority. Among many materials, photocatalytic materials have been emphasized because of their excellent photo-catalytic properties, and they have great potential for environmental protection.[1,2] There are many oxide sulfide semiconductors that can be used as photocatalysts, among which TiO_2 is considered one of the most promising photocatalysts for effective degradation of organic pollutants due to its physicochemical stability, low cost, non-toxicity, and unique electronic and optical properties.[3–8] However, the pure TiO_2 bandgap is wide, the quantum efficiency is low, the compound rate of electron–hole is high, and the visible light utilization efficiency is low, which limits its utilization.[9,10] Therefore, reducing the forbidden bandwidth of TiO_2 and improving the visible light utilization of TiO_2 are of great significance for expanding the practical applications of TiO_2 photocatalysts. It is shown that surface deposition of precious metals, semiconductor composite modification, and ion doping modification of TiO_2 can change the forbidden band width of TiO_2.[11 14] Transition metal and rare-earth metal ion doping of TiO_2 can effectively introduce defects and reduce the compounding rate of electron–hole pairs, while the formation of doping energy levels reduces the forbidden bandwidth and improves the TiO_2 light-absorbing ability.

Compared with powdered and thin-film TiO_2, TiO_2 nanotubes have larger specific surface area and more adsorption space, which can provide higher reaction energy for photocatalytic degradation. Currently, the main methods for preparing TiO_2 nanotubes are template synthesis, hydrothermal synthesis, anodic oxidation, etc.[15] Among them, the anodic oxidation method has the advantages of simple preparation process, simple catalytic conditions, low cost, and highly ordered arrays. In this paper, pure TiO_2, Mo^{6+}, Gd^{3+} mono- and co-doped TiO_2 nanotubes were prepared by anodic oxidation combined with two-step electrochemistry. The selection of Mo^{6+} and Gd^{3+} doping mainly lies in the fact that the radius of Mo^{6+} is similar to that of Ti^{4+}, which can replace Ti^{4+} more easily and realize substitutional doping to make the distribution of dopant ions more uniform, and the doping of Mo^{6+} can form an impurity energy level to delay the electron–hole composite, which greatly improves the performance of photocatalytic reaction. The introduction of rare earth element ions will have an important effect on the formation and lattice distortion of TiO_2 crystals, which will change the crystal shape, grain size, energy band structure, motion state of photogenerated electron-hole pairs, and lifetime of TiO_2. Gd is a rare earth element, and Gd^{3+} has a strong ionic potential that is attractive to electrons and can retard the rate of electron-hole complexation. The microscopic morphology, crystalline structure, light absorption properties, crystalline transition process of different ion-doped titanium dioxide nanotubes have been investigated by SEM, XRD, UV–Vis, DSC, and FT-IR.

2. Experimental Part

2.1. *Experimental materials*

The main drugs and reagents used in this experiment are: titanium sheet (Ti) 99.99% pure and graphite sheet (C) 99.95%. The manufacturer is Hebei Muzhan Metal Materials Co. Methylene Blue ($C_{16}H_{18}N_3ClS$) was purchased from the official website of Sinopharm Reagent, the manufacturer is Sinopharm Chemical Reagent Co. Ammonia Fluoride (NH4F) with a purity of 96%, Ethylene Glycol ($(CH_2OH)_2$) with a purity of 98%, Sodium Molybdate (Na_2MoO_4–$2H_2O$) with a purity of 99%, and Gadolinium Nitrate Hexahydrate (GdN_3O_9–$6H_2O$) with a purity of 99% were purchased from the official website of Aladdin Reagent, and the manufacturer is Aladdin Biochemistry Technology Co. Deionized water (H2O) homemade.

2.2. Sample preparation

A 2×3 cm titanium sheet was used as the substrate, first, the titanium sheet was polished smooth, and the titanium sheet was ultrasonically cleaned with acetone, anhydrous ethanol, and deionized water in turn, and dried and set aside; Ammonium fluoride-deionized water–ethylene glycol electrolyte was prepared at a concentration of 0.25 mol/L. Used the above treated titanium sheet as the anode and graphite sheet of equal size as the cathode. Pure TiO_2 nanotube arrays were prepared by connecting DC constant voltage power supply in electrolyte, adjusting the voltage to 30 v and determining the oxidation time to 1 h.

Next, sodium molybdate and gadolinium nitrate hexahydrate were used as the source of dopant ions, and a certain concentration of aqueous sodium molybdate and gadolinium nitrate hexahydrate solution was prepared as the electrolyte. The dried titanium dioxide nanotubes as described above were placed at the cathode and graphite was placed at the anode for electrochemical oxidative doping of the TiO_2 nanotube arrays with Mo^{6+}, Gd^{3+}, and Mo^{6+}–Gd^{3+}. The concentrations of single-doped Mo^{6+} and Gd^{3+} were 0.005 mol/L and 0.01 mol/L, respectively, and the total concentration of co-doped Mo^{6+}–Gd^{3+} was 0.015 mol/L of which the concentration of Mo^{6+} was 0.01 mol/L, and the concentration of Gd^{3+} was 0.005 mol/L. The doping was carried out at a voltage of 30 v and the reaction was carried out for 1h. Finally, the four groups of titanium dioxide nanotubes were put into a high-temperature sintering furnace and calcined at 450°C for 3 h and then cooled with the furnace, which completed the preparation. These four groups of samples are abbreviated as pure TiO_2, Mo^{6+}–TiO_2, Gd^{3+}–TiO_2, Mo^{6+}–Gd^{3+}–TiO_2.

2.3. Characterization of samples

Samples were analyzed using an X-ray diffractometer manufactured by Bruker D8 AdvanceX, Germany. ZEISS GeminiSEM 300 scanning electron microscope was used to characterize the surface morphology of the samples, with a magnification of 50X~500000X, and an energy spectrometer was used for surface scanning at the same time. The UV–Vis absorption spectra of the samples were tested using a U4100 UV–Vis spectrophotometer from HITACHI, Japan. The samples were tested in infrared by reflectance method using an EQUINDX Fourier infrared spectrometer manufactured by BRUKER, Germany, with a scanning band of 4000–30 cm^{-1}. The thermal analysis was tested by PerkinElmer STA 8000 instrument with air atmosphere, the heating rate was 10°C/min, and the heating interval was 30–800°C.

2.4. *Photocatalytic degradation experiment*

Methylene blue was chosen as the simulated pollutant and 20 mg/L methylene blue solution was prepared. Four groups of prepared TiO_2 nanotube samples were put into methylene blue solution and photocatalytic degradation experiments were carried out under light conditions. The solution was taken every hour and the absorbance at this time was detected by UV–Vis spectrophotometer. Within a certain range, the concentration of methylene blue solution is proportional to the absorbance, and the degradation rate P is calculated according to the formula (1) for the decolorization rate of methylene blue solution:

$$P = \frac{(C_0 - C)}{C_0} \times 100\%, \tag{1}$$

where C_0 represents the concentration after adsorption-desorption equilibrium, and C is the instantaneous concentration of methylene blue solution at each sampling point during the degradation process.

3. Results and Discussion

3.1. *DSC analysis of different ion-doped TiO_2 nanotube arrays*

Figure 1 shows the DSC curves of the four groups of TiO_2 samples. It can be seen that all four groups of samples have a broader heat absorption peak in the 0–80°C section, which is mainly due to the

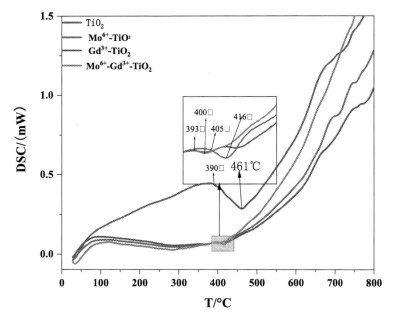

Fig. 1. DSC curves of TiO$_2$ with different doped ions.

evaporation of adsorbed water and free water inside the TiO$_2$ nanotubes. The continuous increase in the heat absorption rate of pure TiO$_2$ samples from 80°C to 390°C is mainly related to the transformation of amorphous TiO$_2$ to anatase by heat absorption. While the Mo^{6+} and Gd^{3+} doped TiO$_2$ because Mo^{6+} replaces Ti^{4+}, trace amount of Gd^{3+} enters into the lattice interstices of TiO$_2$ to cause lattice distortion, and the increase of the internal stress makes Mo^{6+}–TiO$_2$, Gd^{3+}–TiO$_2$, and Mo^{6+}–Gd^{3+}–TiO$_2$ do not need to absorb more heat to lead to the transformation of TiO$_2$ from amorphous to anatase state, so the DSC curves of the doped samples keep the heat absorption rate basically unchanged between 80°C and 393°C. After that, the first exothermic peak occurs by continuing to increase the temperature, for the transition from an amorphous to an anatase state. The first exothermic peak of pure TiO$_2$ is located at 390–461°C, Gd^{3+} single-doped and Mo^{6+}–Gd^{3+} co-doped TiO$_2$ were located between 393–400°C, and Mo^{6+}–TiO$_2$ were between 405–416°C. The transition temperatures of the four groups of samples are all around 400°C, and the phase transition temperature ranges of the doped TiO_2 have become narrower in all of them compared with that of the pure TiO_2. This indicates that the phase transition rate of doped TiO_2 becomes faster. With a further increase in temperature, the rate of heat absorption increases in all four groups of samples, and a second exothermic peak appears, which corresponds to the transition from the anatase to the rutile phase. Pure TiO_2 shows a small exothermic reaction at 670°C, Mo^{6+}–TiO_2 and Gd^{3+}–TiO_2 show flatter exothermic peaks at 710°C and 776°C, respectively, and Mo^{6+}–Gd^{3+}–TiO_2 has a very inconspicuous slowing down of heat absorption at 655°C. The phase transition temperature of single doped Mo^{6+} and Gd^{3+} is shifted to the right, which indicates that the doping of Mo^{6+} and Gd^{3+} increases the phase transition temperature of anatase to rutile and inhibits the transformation of anatase to rutile due to the increase in internal stresses hindering the grain boundary distortion caused by the doping of Mo^{6+} and Gd^{3+}. The transition temperature of Mo^{6+}–Gd^{3+} co-doped anatase to rutile is shifted to the left, and the decrease in transition temperature is due to the fact that the increase in dopant ions provides a site for the phase transition and promotes the transition to rutile, but the presence of lattice distortions hinders the transition to rutile, making the final promotion effect less pronounced.

3.2. Physical phase analysis of TiO_2 doped with different ions

Figure 2 shows the XRD patterns of four sets of TiO_2 nanotubes. As can be seen from Fig. 2, the diffraction peaks of both anatase and rutile phases are present in both pure TiO_2, as well as Mo^{6+} and Gd^{3+} mono- and co-doped TiO_2 nanotubes. Compared with the strongest peak $2\theta = 25.3°$ of anatase in the TiO_2 standard PDF card, the four groups of samples have a larger 2θ at the strongest characteristic peak, which is analyzed to be caused by lattice distortion. $2\theta = 36.1°$ is the diffraction peak of the rutile phase, which corresponds to the crystallographic plane (101). The diffraction peaks at

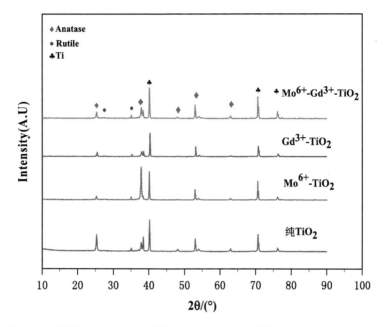

Fig. 2. XRD patterns of different ion-doped TiO₂ nanotube arrays.

$2\theta = 40°$, $53°$, and $70°$ are the diffraction peaks of the titanium substrate. The diffraction peaks of plate titanite were not found in the above four groups of samples, indicating that TiO₂ amorphous were all transformed into anatase and rutile after calcination at 450°C for 3 h. The results showed that the TiO₂ amorphous were all transformed into anatase and rutile after calcination.

From the XRD pattern, the proportion of each crystal type can be calculated according to the Quantitative formula,[16] see Table 1.

As can be visualized from Table 1, the crystal phase structures of pure TiO₂ and TiO₂ nanotubes modified by Mo^{6+} and Gd^{3+} doping are all dominated by anatase, of which the percentage of anatase in pure TiO₂ is 91.6%, and that in single-doped TiO₂ nanotubes doped with Mo^{6+} and Gd^{3+} is increased to 97.3% and 98.1%, respectively, indicating that both ionic doping can inhibit the transformation of anatase to rutile. When Mo^{6+} is doped alone,

Table 1. Calculated results of XRD analysis of different ion-doped TiO_2 nanotube arrays.

Sample name	Phase percentage/%	
	Anatase	Rutile
TiO_2	91.6	8.4
$Mo^{6+}-TiO_2$	97.3	2.7
$Gd^{3+}-TiO_2$	98.1	1.9
$Mo^{6+}-Gd^{3+}-TiO_2$	91.5	8.5

because the ionic radius of Mo^{6+} (0.064 nm) is similar to that of Ti^{4+} (0.068 nm), it is easy for Mo^{6+} to enter the lattice and replace Ti^{4+}, and uniformly distributed in the TiO_2 lattice to form a cosoluble substance, which in turn induces a lattice distortion, accumulates the strain energy, and makes it difficult for the interface to move, thus hindering the occurrence of the phase transition.[9] Gd^{3+} doping, due to the radius of Gd ions (0.0938 nm) is larger than the radius of Ti ions, it is difficult to replace Ti^{4+}, a part of the Gd^{3+} into the TiO_2 lattice interstitials to cause lattice distortions, there are also a number of Gd^{3+} will form oxides distributed in TiO_2 microcrystals to make anatase to rutile transformation difficulties.[17] The larger the difference between the dopant ion and Ti^{4+} radius, the more difficult the phase transition is. The amount of ions introduced increases when $Mo^{6+}-Gd^{3+}$ is co-doped, and only a small fraction of Mo^{6+} and TiO_2 form a solid solution due to the ionic radius, A small amount of Gd^{3+} enters into the lattice interstitials, and the remaining ions are free outside the TiO_2 lattice, which will form compounds or undergo agglomeration to form small clusters distributed around the TiO_2, providing a site for the phase transition and promoting the transformation of anatase to rutile.[18] However, because both Mo^{6+} replacing Ti^{4+} and Gd^{3+} entering the interstitial are inhibiting phase transitions, the proportion of anatase phase in $Mo^{6+}-Gd^{3+}$ co-doped TiO_2 nanotubes is about the same as that in pure TiO_2.

3.3. SEM and EDS analysis of TiO$_2$ doped with different ions

Figures 3(a)–3(d) show the SEM morphology of pure TiO$_2$, Mo^{6+}–TiO$_2$, Gd^{3+}–TiO$_2$, and Mo^{6+}–Gd^{3+}–TiO$_2$, respectively, at 100 kx magnification. As can be seen in Fig. 3, all four sets of samples formed nanotubes on titanium substrates, and the average inner diameter of the tubes was around 70 nm. Figure 3(a) shows the surface morphology of pure TiO$_2$, it can be seen that the nanotube orifice is round, the diameter of the tube is uniform in size and orderly arranged, the surface is flat, there is no collapse, and the overall morphology is very good. Figure 3(b) shows the surface morphology of Mo^{6+}–TiO$_2$, with a regular arrangement of nanotubes and uniform tubing, but localized cracks may be caused by subsequent pinch-holding of the sample. Figure 3(c) shows the surface morphology of

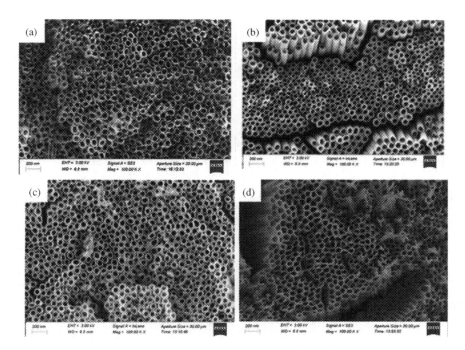

Fig. 3. SEM morphology of TiO$_2$ nanotube arrays with different dopant ions: (a) TiO$_2$, (b) Mo^{6+}–TiO$_2$, (c) Gd^{3+}–TiO$_2$, (d) Mo^{6+}–Gd^{3+}–TiO$_2$.

Gd^{3+}–TiO_2, which is basically unchanged from pure TiO_2. Figure 3(d) shows the topography of Mo^{6+}–Gd^{3+}–TiO_2 compared with pure TiO_2 and single-doped TiO_2, part of the flocculent is adsorbed on the surface, which may be caused by the crystallization of some ions on the surface of TiO_2 with increased concentration of doped ions. From the SEM morphology, it can be seen that the doping of Mo^{6+} and Gd^{3+} basically did not affect the formation and growth of nanotubes, and the nanotubes always maintained a uniform pore size and orderly arrangement, and the most important thing is that the secondary oxidation did not lead to the collapse and destruction of TiO_2 nanotube arrays in the first step of oxidation.

Figures 4–6 show the EDS surface scanning results of Mo^{6+}–TiO_2, Gd^{3+}–TiO_2, Mo^{6+}–Gd^{3+}–TiO_2 nanotubes with EDS surface scanning results. From the EDS results and element distribution diagrams,

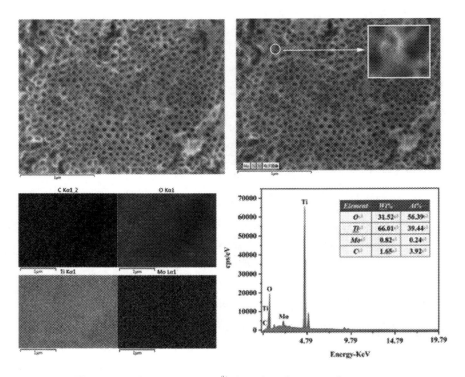

Fig. 4. EDS results of Mo^{6+} doped TiO_2 nanotube arrays.

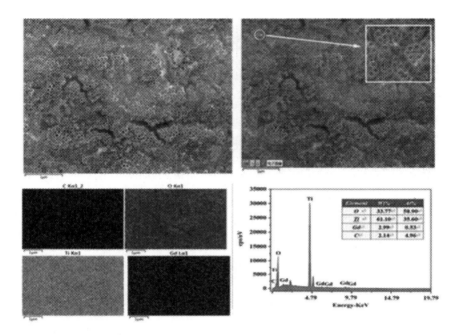

Fig. 5. EDS results of Gd^{3+} doped TiO$_2$ nanotube arrays.

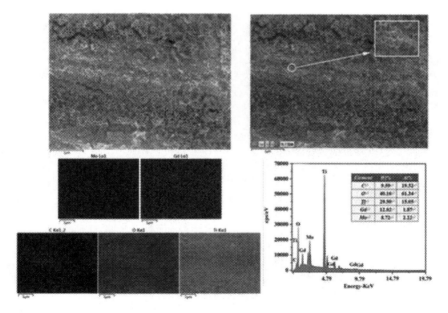

Fig. 6. EDS results of Mo^{6+}–Gd^{3+} co-doped TiO$_2$ nanotube arrays.

it can be seen that Mo^{6+} and Gd^{3+} are able to enter the dioxygenated TiO_2 nanotubes both mono- and co-doped. titanium dioxide nanotubes, and the distribution is uniform. In addition, the high C content on the TiO_2 surface may be due to the adsorption of CO_2 or other organic matter on the sample as well as the preparation of the sample before the test, and the ratio of Ti to O atoms does not show 1:2 because of the high voltage that may penetrate through the TiO_2 scanning to the Ti substrate during the EDS process. The large increase in both C and O compositions in the EDS results of Mo^{6+}–Gd^{3+}–TiO_2 is the result of excessive CO_2 adsorption in the TiO_2 tubular results during sample preparation.

3.4. Ultraviolet–visible spectroscopy of TiO_2 nanotube arrays doped with different ions

The UV–visible absorption spectra of different ion-doped TiO_2 nanotubes are analyzed and the results are shown in Fig. 7.

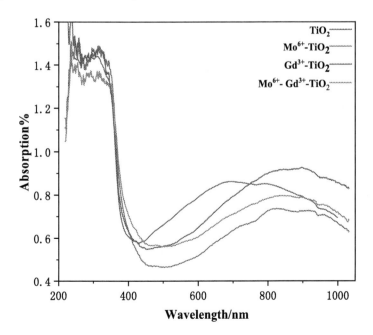

Fig. 7. UV–Vis absorption spectra of TiO_2 nanotube arrays with different ion doping.

Table 2. Light absorption band edges and forbidden bandwidths of TiO_2 nanotube arrays with different ion doping.

Ion doping	TiO_2	$Mo^{6+}-TiO_2$	$Gd^{3+}-TiO_2$	$Mo^{6+}-Gd^{3+}-TiO_2$
λg (nm)	401.2	407.2	403.8	411.4
Eg (eV)	3.09	3.05	3.07	3.01

The absorbance–wavelength curves of the measured samples are processed using the tangent method according to Fig. 7, and the optical absorption threshold λg (nm) is obtained, and then the bandgap Eg (eV) is calculated by Eq. (2),[19] which is listed in Table 2.

$$Eg = \frac{1240}{\lambda g}. \tag{2}$$

As can be seen from Fig. 7, the absorption peak of pure TiO_2 is at 683 nm, and the absorption peaks of Mo^{6+} and Gd^{3+} doped TiO_2 are all significantly red-shifted from near 650 nm to outside 800 nm. Combined with the data in Table 2, it can be seen that the light absorption band edge of the pure TiO_2 nanotubes is 401.2 nm, and the forbidden bandwidth is 3.09 eV, and that the light absorption band edge of the Mo^{6+} and Gd^{3+} -doped TiO_2 all increases, corresponding to narrowing of the forbidden bandwidth. Among them, the change of $MO^{6+}-Gd^{3+}$ co-doped TiO_2 is the most obvious, the optical absorption increases to 411.4 nm, and the band gap widths decrease to 3.01 eV. The main reason for this phenomenon is that Mo^{6+} is a high valence ion, and Mo^{6+} doping of TiO_2 introduces free electrons, which are positively attracted by Mo^{6+} around it, and the energy of this electron is lower than the energy of the electrons in the conduction band of TiO_2, but higher than that of the electrons in the valence band, thus forming the donor levels.[20] Moreover, Mo^{6+} and Ti^{4+} have similar ionic radii, and when doped, Mo^{6+} is more likely to replace Ti^{4+} in the crystal lattice, reducing the distance between Ti^{4+} and O^{2-}, making the band gap of TiO_2 narrower, and the photogenerated electrons are more likely to undergo the transition. Gd^{3+} is a lower valence ion, Gd^{3+} carries three valence electrons, which introduce holes when doped into TiO_2, forming the acceptor level and reducing its forbidden bandwidth.[21] When Mo^{6+}–Gd^{3+} co-dopes TiO_2, there are both donor and acceptor energy levels, and the combined effect of Mo^{6+} and Gd^{3+} makes the energy of the impurity energy level a little bit lower, and the forbidden bandwidth changes most significantly.

3.5. *Performance analysis of photocatalytic degradation of methylene blue*

Figure 8 shows the concentration of pure TiO_2, Mo^{6+}–TiO_2, Gd^{3+}–TiO_2, and Mo^{6+}–Gd^{3+}–TiO_2 nanotube arrays for the photocatalytic degradation of methylene blue solution at different times of light exposure.

As can be seen in Fig. 8, the concentration of methylene blue solution in all four groups of samples decreases gradually with the increase of photocatalytic time, which indicates that all four groups of TiO_2 nanotubes have a degradation effect on methylene blue.

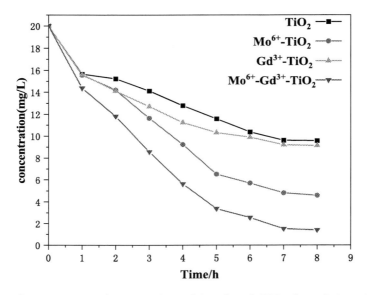

Fig. 8. Concentration of pure TiO_2 and ion-doped TiO_2 degraded methylene blue solution as a function of time.

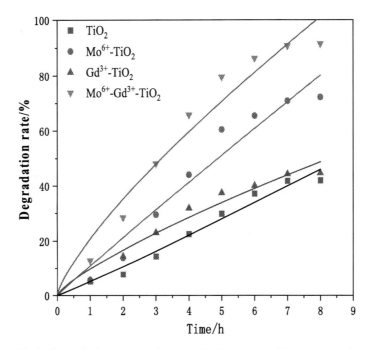

Fig. 9. Variation of photocatalytic degradation rate with photocatalytic time for different ion-doped TiO$_2$.

After 8 h of photocatalytic degradation, pure TiO$_2$ degrades the concentration of methylene blue from 20 mg/ml to 9.5716 mg/ml, while Mo^{6+}–TiO$_2$, Gd^{3+}–TiO$_2$, and Mo^{6+}–Gd^{3+}–TiO$_2$ degrade to 4.5751 mg/ml, 9.1316 mg/ml, and 1.4010 mg/ml, respectively, and the ion-doped samples are all degraded better than pure TiO$_2$.

The degradation rate of methylene blue is calculated according to Eq. (1) and a graph of its variation with degradation time is made, as shown in Fig. 9. From the fitting curves in Fig. 9, it can be seen that the degradation rates of the four groups of TiO$_2$ are basically linear with time, i.e. there is no relationship between the degradation rate and the concentration of methylene blue, and this degradation reaction is a zero-level reaction. The degradation rate increases with the increase of degradation time, and the degradation rate of pure TiO$_2$ is only 41.9% when the degradation time is 8 h. The degradation rates of Mo^{6+} and Gd^{3+} doped TiO$_2$ are higher than that of pure TiO_2, and the degradation rate of Mo^{6+}–Gd^{3+} co-doped TiO_2 is the highest, 91.5%. This is because the outermost electron arrangement of Mo is $4d^5 5s^1$, Mo^{6+} doping will introduce free electrons to form hole traps, constituting the donor level, so that the electrons in the TiO_2 valence band are more prone to transition, the original valence band because of the electron transition and the formation of the hole by the introduction of the free electrons of Mo^{6+} trapping, delaying the compounding of the electron-hole pair. Photogenerated electrons survive longer and more easily reach the TiO_2 surface to react with surface adsorbates. The outermost electron arrangement of Gd is $4f^7 5d^1 6s^2$, the f layer is a half-filled state, more stable, when doped in the form of trivalent Gd^{3+} introduced into the interstices of the TiO_2 lattice. Gd^{3+} has a strong potential energy, can attract electrons around it, delaying the compounding of the electron-hole pairs, thereby improving the photodegradation ability. When Mo^{6+}–Gd^{3+} is co-doped, photogenerated electrons and hole trapping traps are respectively generated in TiO_2,[22] and the synergistic effect of the two makes the photocatalytic degradation effect the best. In the photocatalytic process, trace amounts of rutile play a facilitating role, and Mo^{6+} and Gd^{3+} also make the reaction spectrum of TiO_2 broaden, and the photogenerated electrons and holes on the surface become more, which can produce more photocatalytically active substances and improve the degradation rate.

3.6. *FT-IR analysis of ion-doped TiO_2 nanotube arrays*

In order to investigate the effect of Mo^{6+} and Gd^{3+} on the surface activity of titanium dioxide in titanium dioxide crystals, infrared spectroscopy of pure TiO_2, Mo^{6+}–TiO_2, Gd^{3+}–TiO_2, and Mo^{6+}–Gd^{3+}–TiO_2 nanotube arrays is carried out, and the results are shown in Fig. 10.

As can be seen from Fig. 10, the positions of the absorption peaks of the infrared detection of the four groups of samples are basically the same, and all of them have strong absorption peaks

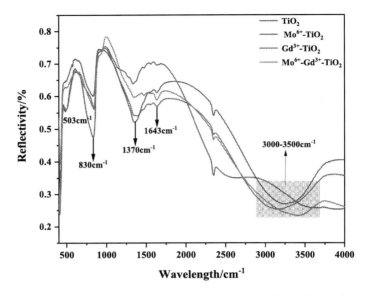

Fig. 10. Infrared spectral analysis profiles of pure TiO$_2$ and ion-doped TiO$_2$ nanotube arrays.

at 503 cm^{-1}, 830 cm^{-1}, 1370 cm^{-1}, 1640 cm^{-1}, and 3400 cm^{-1} wave numbers. The absorption peak at 503 cm^{-1} is caused by the absorption vibration of the Ti–O bond, the absorption of pure TiO$_2$ and Mo^{6+}–TiO$_2$ is the strongest, and the absorption of Gd^{3+}–TiO$_2$ is the weakest; 830 cm^{-1} is caused by the bending vibration of the unsaturated hydrocarbon group =C–H, the absorption peak of pure TiO$_2$ is the strongest, and the absorption peak of Gd^{3+}–TiO$_2$ here is the weakest; 1370 cm^{-1} is the C–H stretching vibration of saturated hydrocarbon group, the absorption of pure TiO$_2$ is weak, and the absorption peak of Gd^{3+}–TiO$_2$ is the strongest. The O–H bond at 1640 cm^{-1} and 3400 cm^{-1} is caused by the O–H bond of free •OH, and the absorption peaks of pure TiO$_2$ are the weakest, and the ion-doped TiO$_2$ nanotubes show significantly stronger and broader peak intensities than those of pure TiO$_2$. The stronger peaks at 503 cm^{-1} for pure TiO$_2$ and Mo^{6+}–TiO$_2$ are due to the fact that Mo^{6+}, Gd^{5+} doped TiO$_2$ makes a small portion of Mo^{6+} replace Ti^{4+} but does not affect the other Ti–O bonds, whereas Gd^{3+}, because of its large ionic radius, is unable to replace Ti^{4+} and can only be found mostly in the interstitial space of the crystal lattice or on the surface of TiO_2, which will form a Ti–O–Gd bond with TiO_2 and lead to the decrease of the Ti–O absorption. The existence of free –OH absorption peaks at 1640 cm^{-1} and 3400 cm^{-1} is due to the fact that TiO_2 generates free electrons and holes and moves towards the TiO_2 surface in the presence of light, the free electrons react with O_2 to form $\cdot O_2$, and the holes have strong oxidizing properties to oxidize organic matter or water adsorbed on the surface to $\cdot OH$. The weakest absorption peaks in pure TiO_2 at 1640 cm^{-1} and 3400 cm^{-1}, and the strongest absorption peaks in Mo^{6+}–Gd^{3+}–TiO_2 are due to the formation of impurity energy levels in doped Mo^{6+} and Gd^{3+} into TiO_2, which reduces the width of the forbidden band of TiO_2 and broadens the reaction spectrum, resulting in more generated photogenerated electrons and holes, and Mo^{6+} and Gd^{3+} provide the electron and hole traps in TiO_2 to delay the electron and hole complexation, thus extending the lifetime of photogenerated electrons, and free electrons and holes which moving to the surface of TiO_2 become more. So it produces more $\cdot OH$, and the $\cdot O_2$ and –OH have strong oxidizing properties. The absorption peaks of Mo^{6+}-Gd^{3+}-TiO_2 are the strongest under the combined effect of Mo^{6+} and Gd^{3+}, followed by Mo^{6+}–TiO_2. Corresponding to the UV–Vis tests of the four groups of samples. The =C–H bond at 830 cm^{-1} and the –C–H bond at 1332 cm^{-1} are caused by the adsorption of organic matter in the air environment and the oxidative reduction of part of the organic matter by TiO_2; from the infrared spectral curves, it can be seen that the TiO_2 is single-doped or co-doped with Mo^{6+} and Gd^{3+} does not lead to the production of other functional groups, but increases the O–H absorption intensity, which indicates that Mo^{6+}, Gd^{3+} doping increases the amount of $\cdot OH$, $\cdot OOH$ adsorbed on the TiO_2 crystal surface, which increases the concentration of highly active groups and makes TiO_2 photocatalytic activity higher. The results of infrared spectral analysis are consistent with the photocatalytic degradation of methylene blue.

4. Conclude

(1) Mo^{6+} and Gd^{3+} doping had an effect on the phase transition of TiO_2, playing a promotional role in the transition from amorphous TiO_2 to anatase, and an inhibitory role in the anatase-phase rutile transition, which increased the transition temperature. Under the heat treatment at 450°C, the anatase composition of pure TiO_2 was 91.6%, and the anatase compositions of both groups of samples, $Mo^{6+}-TiO_2$ and $Gd^{3+}-TiO_2$, were more than 91%, accounting for 97.3% and 98.1%, respectively.

$Mo^{6+}-Gd^{3+}$ co-doping, on the other hand, promotes both the amorphous to anatase transformation and the anatase to rutile transformation; $Mo^{6+}-Gd^{3+}-TiO_2$ has a reduced percentage of anatase composition compared to that of pure TiO_2, which accounts for 91.5%, under the heat treatment at 450°C.

(2) The nanotubes prepared by the two-step electrochemical method with different ion-doped TiO_2 all have an average inner diameter of 70 nm, with flat orifices and effective alignment. The doping of Mo^{6+} and Gd^{3+} does not affect the formation and growth of TiO_2 nanotube arrays, and the TiO_2 nanotube arrays do not collapse during the secondary oxidation, and the surface morphology also maintains the regular arrangement and flatness, and the distribution of Mo^{6+} and Gd^{3+} is uniform in the TiO_2 nanotubes.

(3) TiO_2, $Mo^{6+}-TiO_2$, $Gd^{3+}-TiO_2$, and $Mo^{6+}-Gd^{3+}-TiO_2$ photodegradation of methylene blue were all zero-order reactions, with the fastest degradation rate for $Mo^{6+}-Gd^{3+}-TiO_2$; The photodegradation ability of TiO_2 can be significantly altered by ionic single doping and co-doping, among which the photocatalytic degradation of $Mo^{6+}-Gd^{3+}$ co-doped TiO_2 nanotube array was the best, with its degradation rate reaching 91.51%.

(4) Mo^{6+}, Gd^{3+}, $Mo^{6+}-Gd^{3+}$ doping in TiO_2 crystals constitutes an impurity energy level, which effectively broadens the light absorption band edges of TiO_2, reduces the forbidden bandwidth, and also increases the electrons and holes traps in TiO_2,

and slows down the electron–hole complexes in TiO_2, thus allowing the increase of •OH and •OOH radicals, and improving the photocatalytic activity of TiO_2.

Acknowledgments

This work was sponsored by the Yunnan Science and Technology Talent and Platform Program Project (202105AD160020).

ORCID

J. Du ◉ https://orcid.org/0000-0003-0230-7113

References

1. H. L. Qi, *Energy Conservation & Environmental Protection* **6**, 120 (2019).
2. G. Feng *et al.*, *Appl. Chem. Eng.* **S2**, 213 (2014).
3. H. Y. Li, *et al.*, *Metal Funct. Mater.* **29**, 1 (2022).
4. Y. Han, *Synthesis of zinc oxide photocatalytic material and its photocatalytic performance* (Henan Normal University, Henan 2017).
5. R. Nirmala *et al.*, *Ceram. Int.* **38**, 4533 (2012).
6. C. Gionco *et al.*, *J. Energy Chem.* **26**, 270 (2017).
7. V. N. Rao *et al.*, *Mater. Res. Bull.* **103**, 122 (2018).
8. X. Hong *et al.*, *Chem. Mater.* **17**, 1548 (2005).
9. J. X. Zhu *et al.*, *Salt Indus.* **52**, 6 (2020).
10. F. J. Zhang *et al.*, *Sci. Technol. Bull.* **31**, 6 (2013).
11. D. S. Zhen *et al.*, *Dyeing and Finishing Technology* **40**, 17 (2018).
12. Y. W. Wang *et al.*, *Iron Steel Vanadium Titanium* **41**, 63 (2020).
13. L. Z. Yu *et al.*, *Energy Chem. Indus.* **40**, 11 (2019).
14. M. Faraji *et al.*, *J. Photochem. Photobiol. B Biol.* **178**, 124 (2018).
15. Q. L. Wang *et al.*, *Fine Chem.* **39**, 1106 (2022).
16. R. A. Spurr, *Anal. Chem.* **29**, 760 (1957).
17. L. Y. Su, *Preparation of rare earth element-doped titanium dioxide nanotube arrays and their photocatalytic properties* (Yangzhou University, Yangzhou, 2011).
18. F. X. Li *et al.*, *Mater. Guide.* **9**, 13 (2006).

19. P. Makuła *et al.*, *J. Phys. Chem. Lett.* **9**, 6814 (2018).
20. X. S. Ren, *Jiangxi Chem. Indus.* **37**, 40 (2021).
21. L. Xu *et al.*, *Environmental Protection and Circular Economy.* **41**, 32 (2021).
22. Z. P. Deng *et al.*, *Mater. Guide* **10**, 115 (2008).

Chapter 13

Preparation and photocatalytic degradation performance of AgI/4A molecular sieves

Yaqian Duan ◉ and Jianping Gao ◉*

*College of Science and Technology, Hebei Agricultural University
Huanghua 061100, P. R. China
gaojianping_2007@163.com

AgI/4A molecular sieves photocatalysts with different mass ratios were prepared by depositing the silver iodide on the 4A molecular sieve by ion exchange method. When the mass ratio of silver iodide and 4A molecular sieve was 2:1, AgI/4A molecular sieves exhibited enhancing photocatalytic activity than pure AgI and 4A molecular sieves. The degradation rate of methyl orange (MO) dye reached 96.4% at 20 min. The possible mechanism of photodegradation by AgI/4A molecular sieves was proposed.

Keywords: AgI/4A molecular sieves; photocatalytic degradation; methyl orange.

1. Introduction

In the past time, environmental pollution and the shortage of energy have attracted people's attention. Semiconductor

---

*Corresponding author.
To cite this article, please refer to its earlier version published in the Functional Materials Letters, Volume 16(7), 2340039 (2023), DOI: 10.1142/S1793604723400398.

photocatalysts have the potential to remove pollutants in water and split water to produce hydrogen.[1,3]

TiO_2 and ZnO are common materials in the field of photocatalysis. Due to the electronic structure of ZnO, it can be used as a photo-induced sensitization agent.[4] TiO_2 is widely used because of its advantages of low price, no biological toxicity, strong stability, low energy consumption and no secondary pollution.[5,7]

Due to the wide forbidden band of TiO_2, it can exhibit photocatalytic activity under ultraviolet light. But it cannot be used for photocatalytic degradation under visible and near-infrared light irradiation.[8] Consequently, it is important to develop a new photocatalyst that can be used under visible light (43% of the entire solar spectrum).

Recently, silver-based photocatalysts, such as Ag_3PO_4,[9,10] Ag_3VO_4,[11] $Ag_2Mo_4O_{13}$,[12] and Ag_2CO_3[13] have attracted considerable attention because they could absorb visible light. Unfortunately, the poor photocatalytic stability has restricted its applications. Huang et al.[14] reported that the plasmonic photocatalyst $Ag/AgCl$ is efficient and stable under visible light. Wang et al.[15] reported a hetero-structured $Ag_3PO_4/AgBr/Ag$ plasmonic photocatalyst, which exhibited excellent photocatalytic activity and stability compared to $AgBr/Ag$, $Ag_3PO_4/AgBr$ and pure Ag_3PO_4 crystals. However, the high cost of silver-based semiconductor photocatalysts is not suitable for practical applications. So the development of efficiency and low cost and relative stability AgX-based composite photocatalysts, such as AgI/WO_3,[16] $ZnO@AgI$,[17] $AgI/BiOI$,[18] AgI/SnS_2,[19] Ag_3VO_4/AgI,[20] AgI/Bi_2MoO_4[21] and $AgI/CuBi_2O_4$[22] has become the research focus of the photocatalytic field. Besides, a molecular sieve has been used to disperse the photocatalysts, thereby effectively enhancing its photocatalytic activity.[23,24] However, there have been no reports of precipitating silver iodide on the surface of 4A molecular sieves to improve photocatalytic performance.

In this study, the silver iodide was deposited on the 4A molecular sieve by ion exchange method to prepare AgI/4A molecular sieve photocatalysts with different mass ratios. In addition, the photocatalytic properties were investigated by degrading methyl orange (MO) under visible light. Furthermore, a possible photocatalytic mechanism of AgI/4A molecular sieves was interpreted.

2. Experimental Methods

2.1. *Materials*

4A type molecular sieves, silver nitrate ($AgNO_3$), potassium iodide (KI), anhydrous ethanol (C_2H_5OH) and MO were purchased from Tianjin Kemiou Chemical Reagent Co., Ltd. (Tianjin, China). All reagents were analytical grade and used as received without further purification.

2.2. *Synthesis of photocatalysts*

A sample of AgI/4A molecular sieves composite materials was prepared by a simple ion exchange method. The fine-grind 4A molecular sieve powders were soaked in 0.1 mol/L $AgNO_3$ aqueous solution. Then, the mixture was dispersed by ultrasound for 30 min. The KI solution (0.1 mol/L) was added by drop into the above solution and stirred continuously for 10 min. The obtained product was filtrated and washed with distilled water and ethanol several times. Finally, the precipitate was dried for 24 h at room temperature in air. AgI/4A molecular sieves samples with different mass ratios (1:2, 1:1, 2:1 and 3:1) were synthesized by adjusting the amount of $AgNO_3$ and 4A molecular sieves. The samples was defined as AgI/MS-1:2, AgI/MS-1:1, AgI/MS-2:1 and AgI/MS-3:1, respectively. Pure AgI was also prepared by the same method just without 4A molecular sieves (Fig. 1(a)). The preparation of the photocatalysts was done under no-light conditions.

2.3. *Characterization of photocatalysts*

X-ray diffractometer (XRD, Bruker D8-advance) was used to investigate the crystal phase. Field emission scanning electron microscopy (SEM, Hitachi, S-4800) was used to observe the morphologies. X-ray photoelectron spectroscopy (XPS, Escalab 250xi) was used to analyze the elemental composition. The UV–V is diffused reflectance spectra (DRS, Hitachi U-3010) were used to measure the optical properties.

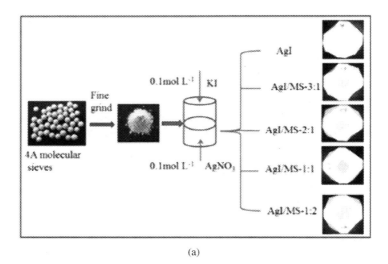

(a)

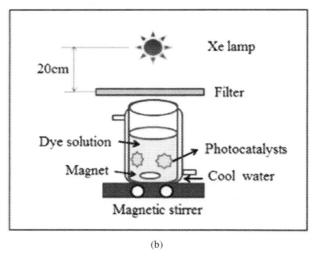

(b)

Fig. 1. (a) Schematic diagram of synthesis process of the AgI/4A molecular sieves samples with different mass ratios (1:2, 1:1, 2:1, and 3:1). (b) Schematic of descending MO in Xe lamp.

2.4. Photocatalytic test

The photocatalytic activity of photodegrading MO was investigated under visible light using a 500 W Xe lamp combined with a cutoff filter (420 nm). Briefly, 100 mg AgI/MS was added into 100 mL

MO aqueous solution (10 mg/L) in a reactor of double layer condensed with running water to maintain a constant temperature. Before light irradiation, the mixture was magnetically stirred in the dark for 30 min to achieve absorption-desorption equilibrium. After light initiation, 5 mL of supernatant was taken at 10 min intervals and centrifuged at 4000 rpm for 5 min. Then, the collected supernatant was detected by UV2550 UV–Vis (Shimadzu) and the maximum absorption wavelength was 464 nm. For comparison, blank experiments without catalysts were performed under the same conditions (Fig. 1(b)).

The degradation efficiency (%) as follows:

$$\text{Degradation } (\%) = (C_0{-}C)/C_0 \times 100\%,$$

where the initial concentration (C_0) was defined as the MO concentration after reaching adsorption equilibrium, and C was the MO concentration at the moment of t.

3. Results and Discussion

3.1. *Characterization of as-prepared samples*

The XRD patterns of AgI, AgI/MS-3:1, AgI/MS-2:1, AgI/MS-1:1 and AgI/MS-1:2 are documented in Fig. 2. The characteristic diffraction peaks for AgI at 2θ of 23.7°, 39.2° and 46.3° are attributed to the (002), (110) and (112) crystal planes of AgI.[17] The XRD patterns of AgI/MS indicate the presence of AgI and MS in the composite.

With the increase of the AgI payload, the relative intensity of AgI characteristic diffraction peak can be increased.

Figure 3 shows the optical properties of the as-prepared samples. It is clear that the 4A molecular sieve has almost no absorption performance, and AgI has good absorption performance in the ultraviolet light region. As can be seen from Fig. 3 above, the absorption performance of AgI/MS samples is constantly enhanced with the increase of AgI content.

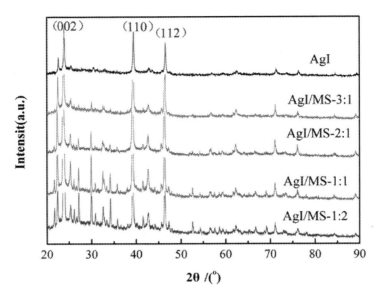

Fig. 2. XRD patterns of AgI, AgI/MS-3:1, AgI/MS-2:1, AgI/MS-1:1 and AgI/MS-1:2 samples.

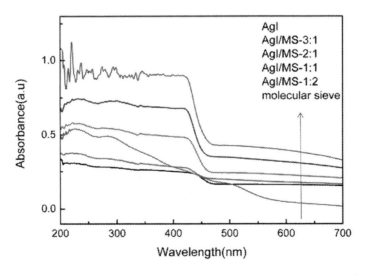

Fig. 3. The UV–DRS absorption spectra of different ratios of AgI/MS.

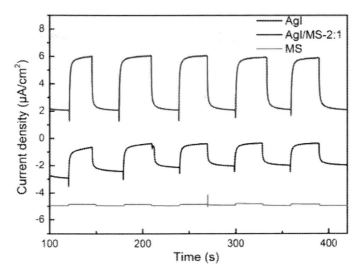

Fig. 4. Transient photocurrent response of AgI, AgI/MS-2:1 and MS.

In the electrochemical experiments, glass slides of different samples were first prepared, which were connected to the electrode in 0.5 M sodium sulphite electrolyte. To study the separation and migration of photogenerated carriers, several cycles of irradiation were performed in 0.5 M sodium sulphite electrolyte. The transient photocurrent response results of AgI, AgI/MS-2:1 and MS are shown in Fig. 4. Thus, the photocurrent of AgI/MS-2:1 is significantly higher than that of AgI and MS. The rapid photocurrent response under visible light irradiation means that photogenerated hole pairs reorganize less and separate more efficiently.[25] Thus, the AgI/MS-2:1 catalyst can more efficiently separate the electron-hole pairs, indicating high photocatalytic activity.

Figures 5(a) and 5(b) present XPS spectra of AgI/MS-2:1 and Ag 3d samples, respectively. The peaks of C, Al, I, Si and Ag of AgI/MS-2:1 sample in the full-scale spectrum could be clearly observed. The fractions of Al, Si, C, Ag and I atoms on the surface and edge are 7.00%, 9.84%, 38.08%, 23.31% and 21.11%, respectively. The Ag 3d XPS spectra of AgI/MS-2:1 is shown in Fig. 5(b).

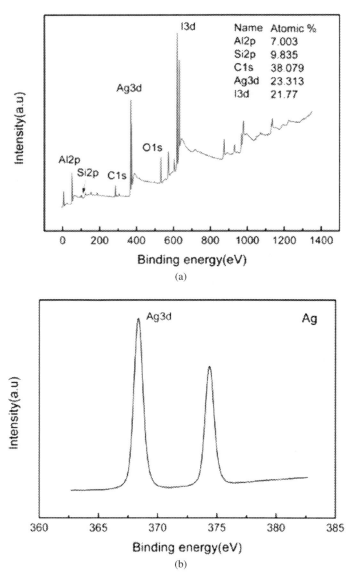

Fig. 5. XPS spectra of (a) AgI/MS-2:1 and (b) Ag 3d.

The peaks at 367.6 eV and 373.6 eV could be ascribed to the binding energies of Ag 3d5/2 and Ag 3d3/2, respectively.

Figure 6 shows the SEM images of the as-prepared samples. The grains of silver iodide are spherical whose diameter is 100–200 nm.

Fig. 6. SEM images of the AgI/MS-2:1.

The 4A molecular sieve particles are cubic, with an edge length of 1–2 μm. As for AgI/MS-2:1, numerous AgI particles have been distributed around 4A molecular sieves and many cubic particles can be observed (yellow marks) in Figs. 6(e) and 6(f).

3.2. Photocatalytic activity and photocatalytic mechanism

The photocatalytic activity of pure AgI, 4A molecular sieves and AgI/MS composites was investigated under visible light irradiation. As shown in Fig. 7(a), 4A molecular sieves could not degrade MO, and the efficiency of pure AgI is less than those of AgI/MS composites. When the mass ratio of the silver iodide and the 4A molecular sieve is 2:1, the decomposition efficiency can reach 96.4% at 20 min. With the increasing degradation time, the color of the MO solution gradually becomes lighter (as shown inset photographs in Fig. 7(a)).

Figure 7(b) shows the degradation rate of MO by different samples after 20 min irradiation. It can be seen that the degradation rate of MO for silver iodide, AgI/MS-3:1, AgI/MS-2:1, AgI/MS-1:1,

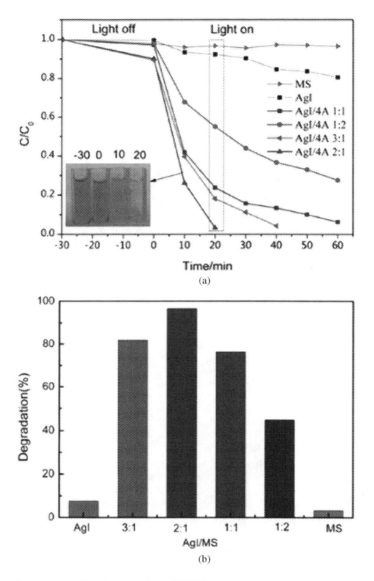

Fig. 7. Photocatalytic degradation of MO by as-prepared samples under visible light: (a) Degradation efficiency and (b) degradation percent comparison at 20 min.

AgI/MS-1:2 and 4A molecular sieve were 7.7%, 81.9%, 96.7%, 76.3%, 45.1% and 3.3%, respectively.

To investigate the stability of the photocatalyst, cyclic degradation experiments were performed for the AgI/MS-2:1. The results are shown in Fig. 8. Figure 8(a) shows the photocatalytic recyclability of AgI/MS-2:1 within five cycles. After visible light irradiation for 60 min, the MO was photodegraded completely in the first cycle. After five cycles, the degradation rate reduced to 64.1%. Figure 8(b) contrasts the photocatalytic degradation efficiency for each cycle at 20 min and 60 min. In the first cycle, the degradation efficiency of MO is 96.4% at 20 min, but the value reduces to 23.3% in the fifth cycle. The reason for the reduced photocatalytic efficiency after three cycles may be that AgI is reduced during photocatalysis; on the other hand, the photocatalyst could be lost during recycling.

Figure 9 shows the photocatalytic mechanism of AgI/MS-2:1 under visible light irradiation. The bottoms of the conduction band (CB) and the edges of the valence band (VB) in AgI are located at about -0.56 V/NHE and 2.24 V/NHE, respectively.[20-22] Under visible light, photoexcited electrons can be transferred to CB from VB of AgI/MS, while leaving equal amounts of holes in VB. Then, photoexcited electrons can be transferred to Ag from CB of AgI/MS. Silver iodide can be excited to generate a light-induced electron–hole (e^--h^+). Ag^+ adsorbed on the surface of the photocatalysts were reduced to Ag. These electrons can be captured by O_2 to form $\cdot O_2^-$. The holes at the VB of AgI can combine with H_2O to generate $\cdot OH$, which can oxidize MO to products. In addition, the molecular sieve can effectively disperse the photocatalysts which can increase its specific surface area and the active sites.

Table 1 shows the photocatalytic degradation capacity of MO by different silver materials. The comparison results show that the degradation activity of AgI/MS-2:1 is higher than that of other silver materials in previously reported papers.

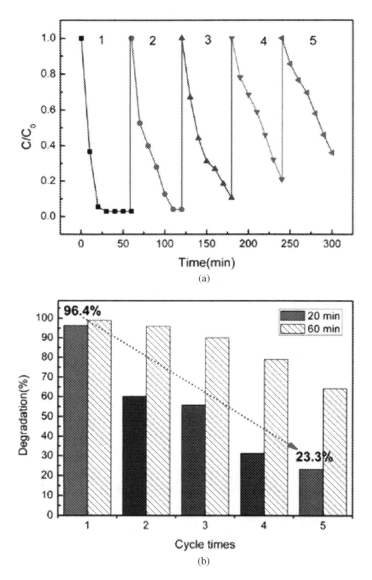

Fig. 8. (a) The cyclic performance of MO removal over AgI/MS-2:1 and (b) photocatalytic degradation efficiency of MO with a recycled AgI/MS-2:1.

Preparation and photocatalytic degradation performance 233

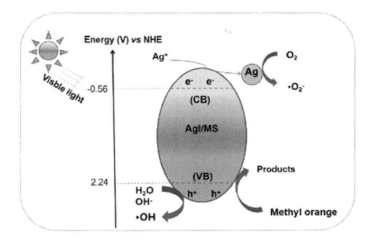

Fig. 9. Photocatalytic mechanism of AgI/MS-2:1 under visible light irradiation.

Table 1. Comparison of the photocatalytic degradation capacities of different silver materials.

Sample	Light source	Target (concentration)	Activity	Ref.
AgI/g-C$_3$N$_4$	Xe lamp	MO (10 mg/L)	DE = 81% (3.5 h)	26
Ag/β-Ag$_2$MoO$_4$	Xe lamp	MO (10 mg/L)	DE = 91.8% (60 min)	27
Ag/AgI/TiO$_2$/CNFs	Xe lamp	MO (10 mg/L)	DE = 97% (3 h)	28
AgI/Bi$_{24}$O$_{31}$Cl$_{10}$	Xe lamp	MO (10 mg/L)	DE = 84.6% (120 min)	29
AgI/Ag$_2$Mo$_2$O$_7$	Xe lamp	MO (10 mg/L)	DE = 80% (90 min)	30
AgI/MS-2:1	Xe lamp	MO (10 mg/L)	DE = 96.4% (20 min)	Present work

4. Conclusion

In summary, a new AgI/MS composite was synthesized by depositing silver iodide on the surface of the 4A molecular sieve by the ion-exchange method. The results showed that the content of silver iodide affected the photocatalytic activity of AgI/MS composites. AgI/MS-2:1 exhibited the highest photocatalytic activity of MO, which can reach 96.4% at 20 min.

Acknowledgments

This work was supported and funded by the Science and Technology Project of Hebei Education Department (No. QN2020265).

ORCID

Y. Duan ◉ https://orcid.org/0009-0004-7825-7233
J. Gao ◉ https://orcid.org/0009-0007-2796-9806

References

1. D. Zhu *et al.*, *Environ. Nanotechnol. Monit. Manag.* **12**, 100255 (2019).
2. S. Sharma *et al.*, *J. Clean. Prod.* **228**, 755 (2019).
3. W. An *et al.*, *Appl. Surf. Sci.* **534**, 147518 (2020).
4. R. Wang *et al.*, *Appl. Surf. Sci.* **227**, 312 (2004).
5. Z. Xiong *et al.*, *Catal. Commun.* **96**, 1 (2017).
6. J. Yu *et al.*, *J. Am. Chem. Soc.* **136**, 8839 (2014).
7. Y. Ma *et al.*, *Chem. Rev.* **114**, 9987 (2014).
8. G. Liu *et al.*, *Angew. Chem.* **120**, 4592 (2008).
9. Y. Bi *et al.*, *J. Am. Chem. Soc.* **133**, 6490 (2011).
10. J. Li *et al.*, *Appl. Surf. Sci.* **372**, 30 (2016).
11. Q. Zhu *et al.*, *J. Phys. Chem. C* **117**, 5894 (2013).
12. Z. Jinfeng *et al.*, *Int. J. Photoenergy* **2012**, 1 (2012).
13. C. Yu *et al.*, *Adv. Mater.* **26**, 892 (2014).
14. P. Wang *et al.*, *Angew. Chem. Int. Ed. Engl.* **47**, 7931 (2008).
15. W. S. Wang *et al.*, *Nanoscale*, **5**, 3315 (2013).

16. T. Wang *et al.*, *Chem. Eng. J.* **300**, 280 (2016).
17. H. Huang *et al.*, *J. Colloid Interface Sci.* **502**, 77 (2017).
18. N. Pourshirband *et al.*, *Chem. Phys. Lett.* **761**, 138090 (2020).
19. Q. Li, *Int. J. Electrochem. Sci.* **15**, 9256 (2020).
20. J. Zhang *et al.*, *Mater. Lett.* **216**, 216 (2018).
21. L. Zhang *et al.*, *Prog. Nat. Sci. Mater. Int.* **28**, 235 (2018).
22. F. Guo *et al.*, *J. Hazard. Mater.* **349**, 111 (2018).
23. Q. Wu *et al.*, *Photocatalyst* **378**, 552 (2016).
24. X. Liu *et al.*, *Sep. Purif. Technol.* **286**, 120400 (2022).
25. X.-J. Wen *et al.*, *Chem. Eng. J.* **383**, 123083 (2020).
26. H. Xu *et al.*, *Appl. Catal. B-Environ.* **129**, 182 (2013).
27. J. L. Zhang *et al.*, *J. Taiwan Inst. Chem. Eng.* **81**, 225 (2017).
28. D. Yu *et al.*, *Appl. Surf. Sci.* **349**, 241 (2015).
29. P. Q. Wang *et al.*, *J. Mater. Sci. Mater. Electron.* **30**, 10606 (2019).
30. Z. Y. Jiao *et al.*, *ACS Omega* **4**, 7919 (2019).

Printed in the United States
by Baker & Taylor Publisher Services